Autonomous and Integrated Parking and Transportation Services

Autonomous and Integrated Parking and Transportation Services

Amalendu Chatterjee Ph.D.

CRC Press
Taylor & Francis Group
Boca Raton London New York

CRC Press is an imprint of the
Taylor & Francis Group, an **informa** business

CRC Press
Taylor & Francis Group
6000 Broken Sound Parkway NW, Suite 300
Boca Raton, FL 33487-2742

International Standard Book Number-13: 978-0-367-18081-2 (Hardback)

Library of Congress Cataloging in Publication Data

Names: Chatterjee, Amalendu (Telecommunications engineer), author.
Title: Autonomous and integrated parking and transportation services / Amalendu Chatterjee.
Description: First edition. | Boca Raton, FL : CRC Press, Taylor & Francis Group, 2020. |
Includes bibliographical references and index. |
Summary: "In this book, the author outlines a Robust Web Parking, Truck and Transportation Portal (RWPTTP) for integrating parking and transportation services - a revolutionary approach in contrast to incremental change for managing traffic congestion. Autonomous vehicle technology, artificial intelligence, internet of things (IOT) and other interconnected hardware and software tools will assist autonomous parking and transportation services and provide next century infrastructure for consolidated transportation customer services. The book highlights currently available autonomous parking and transportation technologies, and the development of an integrated and intelligent transportation service/system (IITS) platform, with specific use of technologies to reconfigure the transportation industry. The author also suggests many regulatory and policy changes to simplify data collection, traffic operation, introduction of duplicate transportation system using light rails (LRs) and High Speed Rails (SPRs) and redistribution of parking spaces along such routes for using renewable energy"– Provided by publisher.
Identifiers: LCCN 2019024845 | ISBN 9780367180812 (hardback ; acid-free paper) | ISBN 9780429059445 (ebook)
Subjects: LCSH: Automobile parking. | Transportation, Automotive. | Autonomous vehicles.
Classification: LCC HE336.P37 C47 2020 | DDC 388.4/74–dc23
LC record available at https://lccn.loc.gov/2019024845

Visit the Taylor & Francis Web site at
www.taylorandfrancis.com

and the CRC Press Web site at
www.crcpress.com

Writing of this book would not have been possible without my doctor wife, Arundhati Chatterjee's constructive criticism of engineering and helpful editing from our two brilliant sons, Avik and Ayan.

Contents

Preface

F ifteen to sixteen million traditional gas fueled cars are being sold in the US every year. If we follow the current minimum requirements for parking space we will add 45 million to 80 million new spaces every year. New parking spaces, in the urban district, will erode prime land for urban housing and other economic development (almost over 0.3 million acres of valuable urban lands lost to parking).[1] If this state continues parking woes will increase as highlighted by Prof. D. Shoup of UCLA in his first book, *The High Cost of Free Parking*. In their second book, *Parking and the City*, Prof. Shoup and his colleagues articulated very well the problem of having too many parking spaces (lack of planning). They recommended elimination of off-street parking (other-wise known as PARCS) in favor of on-street parking with fair prices. In their arguments, smart technologies and their full exploitation did not get full attention. Future technologies will impact parking and trans-form vehicles from gas fueled to purely electric. In other words, society will face a new challenge of managing autonomous parking and trans-portation services – last mile. Vehicle technologies will transform all vehicles (cars or trucks) to be autonomous and electric as well. A common solution of transporting goods and people for the last mile can take a new prominence.

That is why, in this book, I bring technology power to the fore for full consideration by all stakeholders. In my humble opinion, neither of the books mentioned goes far enough for a real technology-based solution that may improve costs nationwide. Parking spaces (requiring private and public consolidation) are to be considered as national resources and must be raised at that level for a standardized uniform solution. Autonomous

---

1 Richard Florida, 'Parking Has Eaten American Cities', *NYT*, July 24, 2018.

vehicles, the iCloud platform, big data analytics (BDA), artificial intelligence, no meter and no gate, sophisticated software instead of field hardware, a parking database, personal devices such as smartphones with smart APPs, autonomous identification, and autonomous invoicing will dictate the future parking paradigm. End users like airline passengers do most of the work to get their wish granted – via a one click solution. Parking solutions should be the tail end (last mile) of other integrated virtual transportation services. I contend that the process (no meter and no gate) I propose will address the above authors' third theory, 'Spend the Parking Revenue to Improve Public Services on the Metered Street' differently – revenues from new services and conveniences called the new paradigm of smart parking. I rationalize other details to combat their arguments starting with renewable energy as applicable for transportation and parking services.[2] In this context, I raise the idea of renewable energy for open parking spaces in accommodating electric vehicles (EVs) and energy self-sustainability. Other modern technologies that integrate total transportation services may also be considered so that how we park becomes a national pride not a daily drag. State and local legislators must be proactive in introducing progressive transportation policies.

Yes, there are innovations, technologies, and also increased costs and congestions. Main reasons for problems include:

- No scientific process is in place.

- Road centric transportation focuses only on the city business center (CBC).

- There is no workforce development through education and training.

- Vendor specific products have no open interface for interconnectivity.

- There is a monopoly attitude and reluctance overhaul due to political and financial influence.

- The parking bubble isolates those concerned from facing the reality of other transportation services for convenience and experience.

---

2 In May, I wrote an article (*Triangle Business Journal*, North Carolina) on renewable energy legacy. Jim Warren of NC Warn (*News & Observer* article, August 6, 2018).

The good news is that all experts agree a change is needed and have been researching the subject. In parallel, there is also a dilemma about how to proceed. There is an existing infrastructure with embedded standalone orientation of all parking spaces – on-street and off-street. One group of researchers argues in favor of incremental change, using the existing infrastructure. They are afraid of too much disruption of their embedded base that may impact the bottom line drastically. I am in favor of radical changes. The rationale for radical changes is as follows:

- The industry is big enough to absorb such change (over $30 billion).
- The number of cities and their populations are growing noticeably (more city living).
- City land must be saved from new parking development.
- Telecommunication industries, airline industries, and software industries (IT) have gone through similar radical transformation.
- Artificial Intelligence may bring another new orientation.
- We can learn from successful disruption technologies.
- Growth and interconnectivity of many transportation services can come under one umbrella.
- A duplicate transportation system (Light Rail, LR, and High Speed Rail, SPR) can be introduced along with road only transportation connecting cities and regions.
- Autonomous vehicles for virtual transportation can be included.
- Greenhouse gas (GHG) emissions can be reduced.

The author presents a new infrastructure that is flexible and open. The major characteristic is accommodating the old embedded base into the new one rather than fitting new elements into the old infrastructure. New features, functions, and terminologies are defined for the benefit of all stakeholders and also wide demography. Examples are:

- There is no difference between on-street and off-street parking.
- All parking spaces are open lots ready with Electric Vehicles (EV) plug-in.
- Open lots are away from the CBC in a newly defined suburban parking district (SPD).
- A trucks parking district (TPD) is defined to include them in the same infrastructure.
- SPDs and TPDs are distributed along duplicate transportation system such as LR and SPR.
- Open lots may be equipped to include solar panels for sustainability with renewable energy.
- Economies of scale are on two fronts: minimizing rural/urban disparity and common and uniform technology/product development across all transportation systems.
- Mobility, software, and artificial intelligence will be built-in capabilities in all devices for remote monitoring, service provisioning, and security.
- Going paperless and pay as you use provisioning are incorporated.
- Drones used for autonomous enforcement with no human bias.
- There is collection of online data for future planning and instant service development on demand.
- Training courses on consolidation of an integrated transportation system are possible for workforce development in community colleges as well as universities.

The new infrastructure has been named the Robust Web Parking, Truck and Transportation Protocol (RWTTP). The ideas presented here have enough materials for all stakeholders to think through. Hopefully, policy framers, designers, and decision makers will make the right decision to follow through. The author's analysis suggests the rate of return on investment (ROI) will be shorter than expected after the maturity of all technologies presented in the book.

About the Author

A Nortel, GTE and Fujitsu alumnus, **Dr. Chatterjee** has done pioneering work in the area of packet networks, ISDN, digital switching, ATM, network connectivity standards, and wireless technology. He brings over forty years of expertise in specific knowledge of the information technology and telecommunications field, associated with business planning. He was Director of Global R&D for Fujitsu and published technical papers. He led a group in the strategic business rationales for creation of the first broadband superhighway in North Carolina using ATM technology for distance learning and telemedicine applications. Since leaving Fujitsu, Dr. Chatterjee has acted as a technical and business advisor to Network Optic Communications and Quad Research. He also co-founded Eximsoft International and has been running it successfully for the last fifteen years. He has pioneered a prototype development of secured mobile e-commerce following the Mobile Electronic Transactions (MeT) standard (at its early stage in 2000s) in collaboration with Ericsson (mobile company) and Entrust (an internet security company) that was showcased in many countries. He has two trademarks (TPP^{TM} and SMA^{TM}) on new business ideas. He earned his Ph.D. in Computer Communications Networks at the University of Ottawa, Canada.

Thinking of a New Parking Paradigm

INTRODUCTION

There is a wise saying: 'Careful Thought + Loving Action = The Power to Withstand Pressure'. For the last twenty years, I have developed a passion for the parking industry against all norms or legacies. I have published many thought-provoking articles in different parking journals and this book is a compilation of those articles.[1] It all started with a project on 'Mobile e-Commerce' with Ericsson. Ericsson's mission was to go beyond silly cell phone games, a non-sustainable commercial venture. Three companies shared their visions of a killer application for the cell phone – mobile transactions for the parking industry with a one click solution. Materials articulated in this book result from careful thought and loving action by the author on emerging technology. The author has also been promoting a national parking standard to bring a uniformity to the industry along with different transportation services. A standard is also required when we talk of connectivity of technologies with mobility to derive full benefits. It is up to the readers to bring about something noble while implementing this new parking paradigm.

---

1 *The Parking Professional Magazine, Parking Today, National Parking Association Magazine, Traffic Infra Tech* and different parking conference papers.

The parking industry's woes are well known.[2] In spite of this, empirical formulas are used to support building more parking spaces. Each new car sold in the industry adds over three more parking spaces, eroding the city's prime land. More than 17 million traditional gas-powered vehicles were sold in USA in 2016. You can now imagine the wastage rate of our prime land. Electric and autonomous vehicles are also progressing strongly in the market. Such vehicles and interconnected mobile devices will need radical thought on autonomous parking and transportation services. Other technological and application trends will also be influential. One needs to understand technology, the relevant terminologies and influence of social media to make an informed decision. The author proposes a next-century national web-based parking platform (WBPP) like Google, containing all relevant information for friendly, consolidated and customer-focused transportation services, even if it means getting rid of cash oriented on-street meters and off-street gates (Parking Access Revenue Control System – PARCS). Those machines are mostly mechanical and labor intensive. They operate on strict, outdated, and biased enforcement policies. They cannot be web enabled. We need customer oriented new policies for parking services like other transportation services via Priceline.com, Expedia.com, etc. Web-based transportation infrastructure for the last mile for transporting goods and people that will include parking provides many advantages. Various technologies and associated terminologies have influenced my thoughts. Each of them, standalone or in combination with others, will play a key role in revolutionizing the consolidated transportation industry. For a century, American family life with two adults has needed at least two cars to get around. Ride-hailing services provided by Uber/Lyft and the demography of the millennial have been reversing the trend. Now, the first mile and the last mile of the major transit station or HUB have become the center of attention for many auto and truck manufacturers: energy efficient and self-driving/self-parking with artificial intelligence (AI) and sensors. Smart software, software emulation, application program interface (API), and big data driven implementations are taking center stage. Knowledge-based information, such as parking spaces and their availability, vehicle identification number, owner address and license plates, etc. will be customized in a data-driven programming

2 Donald Shoup, *The High Cost of Free Parking* (Chicago, IL: American Planning Association, 2005). Donald Shoup, *Parking and the City* (New York: Routledge, 2018).

environment to create new revenue generating services. A clear explanation of the terminology is necessary to set the context in relating emerging technologies to a fully automated, online and instant service through WBPP or the robust web parking, truck, and transportation portal (RWPTTP). RWPTTP may be defined as a next parking Google with relevant interfaces for all parking stakeholders (private and public), and operators. It will allow them to generate income from well-coordinated revenue streams proportional to their resource allocations following the virtual business model of Abnb. Mobility, Big Data Analytics (BDA), or Big Data Mining (BDM). A web approach will also address customer concerns instantly with friendlier interfaces.

Many lessons learnt from airline, telecommunication/information, and software industries could be applied here. Major ones are workload and responsibility sharing with the customer, resource optimization and mobility, and hardware emulation by software and artificial intelligence (AI). For example, airlines have reduced overheads by letting customers buy tickets online, self-check in, printing boarding pass, etc. Telecommunication/information technologies have optimized resource sharing with multiple clients (iCloud) and also extended the remote reach (mobility). Software emulation of many field hardware functions may serve as a good business model for the future success of the parking and transportation industry's new paradigm – reducing costs and customer conveniences. Chapter 3, 'Technology Landscape and Interconnectivity', previews technologies and associated terminologies that may support my hypothesis.

Historical Perspectives

Writing a book on the parking industry, focusing emerging technologies and consolidation of transportation services for the public in the twenty-first century, is very timely. Our nation's cities are becoming more heavily populated with vehicular traffic. The result is overparking. Technologies, innovations and environmental hazards are evolving so fast that a revolutionary approach in the parking industry is required to enable market dynamics (competitiveness) and service dynamics (user friendliness) to match those of other non-transport or transport industries. Multiple factors such as social media, autonomous vehicles, electrical vehicles, intelligent and integrated transportation services (IITS), Generation Y, ride sharing for the last mile, traffic congestion, and greenhouse gas emission will play a significant role shaping the future. We can think of multiple reasons why the on-street meter (over seventy-five years old) and PARCS (over sixty years old) require new thought for modernization.

First, technological advances in the parking industry alone have not met user expectations. For example, latest street meters (pay by phone or pay and display meters) are overly sophisticated, expensive, and are too space specific to conform to modern social norms – 'pay as you use' irrespective of location with a time stamp. Similarly, PARCS, due to 'stop and go' or so-called access gates, slows access in the garage, causing congestion. Not all PARCSs are flexible or web enabled. In addition, the system costs money and labor to operate. Both may be eliminated with alternative provisioning.

Second, parking infrastructure is predominantly operated through legacy systems that have been developed based on hardware and regulations that both need an update to be a market player of the twenty-first century. More effective user-friendly and environmentally conscious options exist today that can be explored to save costs and enhance customer expectations or convenience. Just replacing the hardware may not be a wise idea because of continuous changing technology dynamics such as smart software, iCloud, and internet connectivity.

Third, the parking industry is ripe with economic potential world-wide in addition to providing parking services. The industry is worth over $100 billion annually.[3] In the US, it is close to a $30 billion industry comprised of on-street ($9 billion per year) and off-street ($21 billion per year) market. An overhaul of the parking infrastructure would allow customers to save time while looking and paying for parking spots. This would enable a higher rate of parking turnover, a more efficient allocation of parking spaces, and create a higher revenue generation for parking providers. In addition, many peripheral services can be added at the parking spot (vehicle detailing) or on the website (real-time marketing advertisement).

Fourth, municipalities are not always keen on rapid change or to adopt technologies for public benefit. Political consensus among vested bodies prioritizes short-term budgets over long-term benefits. Administrative bodies are reluctant to rock the boat when stakeholders and the

3 Aashish Dalal, 'The Future of the $100 Billion Parking Industry', Pando.com, January 30, 2014.

public interests are not in sync. Such thinking can be changed only when new policies and benefits are debated and enumerated with a technological overhaul.

Fifth, a number of Intelligent Transportation Systems (ITS) applications are market-ready and could offer convenience, cost reduction, increased revenues, and parking lot, security especially around airports and other key government installations. Trusted Parker Program (TPP™) is such a mechanism introduced by Eximsoft International.

Sixth, a parking overhaul could catalyze a smartphone or web solution for all parking needs, such as availability of free space, rate information, and direction to get there. A MapQuest like solution for parking to guide drivers to an available parking space is a possibility.[4]

Seventh, parking spaces should be listed as national resources as they occupy prime land, especially in major cities. They are mostly segmented. Without a consolidated online view, there can be less than 30% utilization in many locations. A national and international standard to develop such consolidated view is required to bring uniformity in the parking process when technological upgrades are considered. Such technologies can enable less-developed countries to leapfrog market development, as in the mobile phone, to revolutionize the parking industry without going through the transitional plan of the embedded base.[5] It is encouraging news that the International Parking and Mobility Institute (IPMI) has undertaken such efforts with their European counterparts.[6]

Eighth, parking spaces are important to both the private entity and the public entity. Sectoral interests and tradeoffs for business, government, and individuals should be considered. For example, unnecessary parking spaces could be converted to other uses to conserve land and contain costs. Business and housing developments, including recreational aspects of those freed up spaces, cannot be ruled out.

Ninth, competitive ideas and business models developed with digital and software technology in the telecommunications industry, the airline industry, and the latest transportation and hospitality industries like Uber/Lyft and Abnb may all be applicable to parking services. Lessons

---

4 It is a trademark of Eximsoft International.
5 Sam Pitroda and Mehul Desai, *The March of Mobile Money: The Future of Lifestyle Management* (New York: Harper Collins Publishers, 2010).
6 'IPI Alliance Data Webinar', September 25, 2018.

can be learned of software power, routine work, load sharing responsibilities with customers, and online real-time monitoring of the network status. Relevant unique components of other industry models can be borrowed to enhance the parking industry.

Last, but not least, materials in the book are consolidated for both learning and instruction. This book could be used not only to raise awareness about our current parking deficiencies and areas for improvement but also to develop course offerings and trainings to encourage future innovations. Course offerings and trainings on drone technology at Stamford University may serve as a model[7]. The 9/11 terror attack and online security concerns may provide additional research ideas for refinement.

The Bibliography to this chapter highlights a list of publications and conference presentations, including the results of research, for over eighteen years. The initial motivation was to improve the airport parking area security. An article, 'Enhancing Parking Security Against Terror Threats', published in the *Parking Professional Magazine*, May 2004, was the starting point. Unfortunately, the Transportation Security Agency's (TSA's) obvious priority was not the security of parking spaces around airports, seaports, railway ports, or any key government installations – a surprise to many of us. Their prime priority was the security of the airport terminals, passengers, and planes only – a very narrow focus though many parking spots are very close to the runway and may be susceptible to a car bombing or a suicide attack as had happened in Oklahoma and World Trade Center garages.

Future Outlook

Once Nelson Mandela said, 'It always seems impossible until it's done'. It is an enormous task to articulate a vision of the future in the face of the established legacy. *Autonomous and Integrated Parking and Transportation Services* accepts the challenge.

Prof. D. Shoup was the first author to highlight parking problems in big cities in his first book, *The High Cost of Free Parking*. He and his other colleagues articulated very well the problem of having too many parking spaces (lack of proper planning data) and a solution approach

---

7 *The Center for Internet and Society at Stanford Law School is a leader in the study of the law and policy around the Internet and other emerging technologies.*

in his second book, *Parking and the City*: elimination of off-street parking (otherwise known as PARCS) in favor of on-street parking with fair prices. Both of his books have been accepted by the parking community as a benchmark for improving the industry. In addition, many articles in the *Parking Professional Magazine*, the *National Parking Association (NPA) Magazine*, and *Parking Today* shed light on different aspects of the industry. As indicated earlier, every new traditional vehicle adds three to five parking spaces in the urban district, eroding prime land for urban housing and other economic development.[8] On-street parking is biased towards the rich who can pay high prices and fines. Daily on-street and off-street parking competition leads to vehicles cruising for open spaces. In my opinion, none of the solutions goes far enough or is given national importance. A standardized uniform solution must be found. I contend that the process (no meter and no gate) I propose will address their third theory, 'Spend the Parking Revenue to Improve Public Services on the Metered Street' differently or more elegantly for more revenues with many new services. In an article, 'No Meter and No Gate: Auto ID Like Automated Toll Services?', published in *Parking Today* (October 2018), I rationalize other details to debate their arguments. In May 2018, I proposed 'A Legacy on Renewable Energy in NC' in an article in *Triangle Business Journal*, North Carolina. The same idea was highlighted by Jim Warren of North Carolina Warn (*News & Observer* article, August 6, 2018). In my hypothesis, I extend the idea of renewable energy for consolidated open spaces in accommodating electric vehicles (EVs) and energy self-sustainability. Other modern technologies that integrate total transportation services may also be considered so that how we park becomes a national pride not today's daily drag. This will also help state and local legislators to be proactive in introducing progressive transportation policies in addition of cash incentives to attract big corporations such as Amazon, Apple, etc. to North Carolina or other states. My article, 'A Fully Automated Future Transportation System Surrounding Research Triangle Park (RTP) – A Case Study', published in the *Parking Professional Magazine* (November 2018), describes the strategy in detail.

Innovations and technologies have revolutionized our social fabric, disrupting established industries. 'First in Flight' (the Wright brothers)

---

8 Richard Florida, 'Parking Has Eaten American Cities', *NYT*, July 24, 2018.

is one example, and it is a source of pride for North Carolina. Ford's motor vehicle is another. Going to space is an exemplary one. The telecommunication industries used wireless and softwas to contain costs and promote ubiquity with mobility. Then came the information technology (IT) revolution and multi-media services (social media as it is known). In the information age, industry disruptions happen much faster (within just a few years) than in the past. Who thought drones could disrupt the tire industry for transportation services of the future? Electric vehicles will disrupt the fuel industry (gas stations). Self-driving and self-parking vehicles are on their way to disrupt taxi services. Millennials' car ownership and driving habits will be different. It is not inconceivable that the Department of Motor Vehicles (DMV) may go paperless with electronic license plate displays and one must look at the parking industry from this perspective.

On-street and off-street parking though dominates the parking industry's earnings are labor intensive, time consuming and hardware oriented requiring field maintenance and cash collection. Strict and unfair enforcements give an unfriendly image of the industry. Of course, over the years these hardware devices have become very sophisticated but expensive. Mobile and credit card payments have been added for incremental values but not at the level modern technologies can provide. The upgrades needed include a national parking database for proper counting of resources, private and public parking space coordination, integration with other transportation services such as train, airline, hotel, car rental, etc., synchronization of parking services with other virtual transportation services such as Uber/Lyft, and driverless transportation services. We need to think of transportation services in parallel to road transportation, while parking spaces are totally reoriented along new LR and HSR. Such a duplicate system will equally address rural and urban economic and educational development while deploying new automated new technologies.

Similar arguments are also true for 15 million trucks transporting goods across Interstate Highways but not having enough parking spaces on the Interstate Highway Corridors. These parking spaces and other natural resources lend themselves to an opportunity of a possible technological legacy for the public to appreciate nationwide. Parking spaces being national resources, way must be found for optimum reorientation. Along with it will come a visionary parking infrastructure like Google (search engine) for all vehicles (cars, buses and trucks) linking other transportation services for goods and people. Let private and public spaces form a database for online information such as rates, occupancy, availability and routes to get there – a PARKQUEST™ like MAPQUEST.[9] PARKQUEST™ works for parking space as MAPQUEST (Google Maps) does for going from place A to place B. Institute of Parking Industry (IPI), the biggest parking industry association in the US, is collaborating with European parking communities for a similar vision of a national standard for uniformity. A new approach is required. A suburban parking district (SPD) may be defined to relocate all on-street and off-street parking spaces from CBC to SPD so that people can walk freely with no car cruising congesting the city street. Automated transport vehicles should transport people from these SPDs to the city center. This is the first step to unite on-street and off-street applications under one umbrella creating a larger pool of open space. This also rationalizes the second step to eliminate meters and gates to restore city land for urban housing and other economic development. SPDs being located in a convenient place (mostly rural boundaries) away from the city center will also help to reduce rural and urban economic gaps. All parking spaces will be available for online reservation remotely anytime and from anywhere for one click solution – similar to locating a supercharging station for Tesla EV. Additionally, auto location identification and auto invoicing via latest enhanced geo positioning systems (GPS) will simplify the billing without even any reservation. Along with it, 'Vision Zero' implementation will reduce road accidents and other injuries – a Scandinavian mission.

Concepts such as Suburban Truck Parking District (STPD) can be defined in the Interstate Highway corridors in coordination with weighing stations. Many benefits of the above could be derived for transporting goods. No truck will enter CBCs or even highways in local areas avoiding delay for commuters during office hours. Instead, automated vehicles (smaller trucks) or drones will be used for delivering goods door to door – last mile delivery. A parking arrangement for drones may also be needed in this newly defined STPD. These STPDs are in rural areas creating an opportunity for high paying jobs for local people. Goods transportation and managing national and international (Canada and Mexico) 15 million truck journeys using our highways across the country

---

9 A trademark adopted by Eximsoft International.

can be fully automated if progressive policies are adopted. Truckers can upload their routes, stop-overs, load types, their maintenance and driving records as they travel from start to finish. Truck drivers can also report road conditions and fatalities remotely for highway safety. In addition, resources at the weighing station close to these TPDs can be planned in advance depending on the truck regulations and their arrival at the station.

The above ideas can be combined with a simple Internet or virtual infrastructure interconnecting devices using an iCloud platform and software emulation of hardware functions as much as possible. The iCloud platform is getting smarter everyday with artificial intelligence (AI) tools to make statistical decision. Wireless for mobility and data analytics (big data mining) for accuracy (data generated by all connected devices as required by regulations) will form the backbone. It will be like a concert where all instruments play in harmony. Here in the transportation concert, diverse technologies will play in harmony to connect or disconnect devices as and when needed. Geo positioning system (GPS), Bluetooth beacon, Bluetooth low energy emission (BLE), PV cells, EVs, drones, cars, trucks, etc. playing in harmony connected via Internet of Things (IOT) for transporting people and goods[10] shown in Figure 1.1 – future world of connectivity. Benefits derived from the above orientation will be numerous, such as the configuration and reconfiguration of parking spaces to increase revenues on a short notice, synchronous transportation and parking services, consolidated private and public parking spaces for efficient land use, self-identification of drivers and vehicles for auto invoicing, auto identification of valet and handicapped drivers bringing uniformity with improve security. Every 500 such open spaces could potentially generate electricity (2.5 MW) for charging EVs and other use – self-sustainability. In addition, the process reduces air pollution. The greatest benefits are reducing revenue collecting hardware costs by 1/20th ($800 per space to $40 per space) and opening one click solution. In fact, the payoff period of new technologies may be shorter than any other existing business model. In addition, civility in the parking industry can be restored where uncivility has been reigning since the start of the auto industrial revolution. Civility should not cost extra money rather it should save money in the long run. One needs to go beyond the short time gain and

---

10 Internet Wikipedia on IOT.

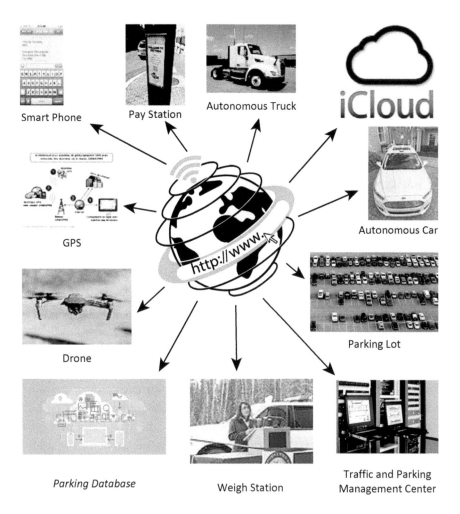

FIGURE 1.1 Autonomous (RWPTTP) parking – technology connectivity outlook.

bring social uplifts and long-term benefits to all stakeholders. Restoration of civility is also important to reduce the use of judicial system resources. A civil environment can be restored by different ways. Online automations, hardware minimizations, no manual spot penalties and collections, workload sharing, unbiased regulations, sharing responsibilities, instant report generation and conflict resolutions with time stamped digital records are elements to orchestrate civility. Increased revenues through new services and defining a new regulatory framework called 'Affordable

Integrated Transportation and Fair Parking Act (AITFPA)' will also bring the desired result. Critical examination of the above to show the saving in both capital investments and operation costs with new proposed infrastructure is essential. Instant rate changes with broadcast capabilities based on market dynamics will be rationalized as opposed to having arbitrary increases. Yes, many radical changes in the parking industry are proposed for readers to think through. There are so many possible variables to accommodate. Always remember, as the technology futurists Dan Sullivan and Carl Richards note, 'The moment your past becomes bigger than your future, you die'. The author will be highly pleased if materials presented here can make useful dents in the mindset of industry leaders, and social, economic and political thinkers and decision makers.

Why a Revolutionary Approach?

Business as usual or incremental change as has been done over the last several years is not sustainable for the twenty-first-century parking industry in isolation. In addition, it may lead to a national environmental disaster. That is why I propose a measurable and achievable approach that is possible if we consider a holistic approach with the available and the on-the-horizon technologies. Such approach will last for several decades – till the end of twenty-first century. With over sixteen years of research, I am convinced more than ever that the parking industry will need a revolution because it is a late player in the game compared to similar customer service industries of today. Here are some of my additional research findings to justify the paradigm change.

Image Reflection

The parking industry has to remake its image and continue to evolve to satisfy customer demands. Other service industries have managed to progress and are much further along with many advancements. Meanwhile, the legacy parking systems are still unable to synchronize with the government initiated 'ITS Strategic Plan – 2015–2019' developed by the Joint Program Office (JPO). A revolution is required to meet ITS capabilities and raise parking services to the next level of customer satisfaction. The first step of the revolution is to transform the parking industry from the hardware configuration to a software configuration – web enabled nationwide. The second stage of revolution is to enable parking applications to be written by third parties so that they can be easily integrated with base services instantly with no delay associated with hardware installations.

Cost/Rate Disparity

According to Professor Shoup, the current parking platform has increased a wide disparity between the rich and the poor. He stated:

> The federal and state governments give cities billions of dollars every year to build and operate mass transit, yet most cities require parking based on the assumption that everyone will drive everywhere. The uncoordinated actions of many individuals can add up to a collective result that most people do not like. In this case, the uncoordinated parking requirements of many cities can add up to an asphalt wasteland that blights the environment and decants transit riders into cars.[11]

He further suggested, 'We can achieve enormous social, economic and environmental benefits at almost no cost simply by subsidizing people and places, not parking and cars.' Prof. Shoup has recommended an incremental paradigm shift, quoting the famous Princeton science historian, Thomas Kuhn. I propose a revolutionary approach with a better and efficient utilization of all parking spaces or even reducing total parking requirement policies when they are all consolidated in key locations away from the city business center (CBC) with online 'pay as you use' pricing – a theory of larger and a consolidated pool of resources to share.

Fair Market Price for Curb Parking

Prof. Shoup introduces fair pricing for street parking to preserve high land use compared to off-street parking. Currently, cost structures for on-street and off-street parking favor on-street parking, because the price point for on-street spaces is kept artificially low for on-foot commerce and recreation. In the revolutionary approach, all parking spaces (on-street or off-street) are removed from the CBC. Such removal along with the elimination of meters and gates will bring an equilibrium in the market place and also uniform business applications.

Use of Increased Revenues for Public Improvements

Increased revenues from Shoup's fair pricing can be used to improve many aspects of city living. For example, municipalities can use these

---

11 Wikipedia – Territory Education Facilities Management Association (TEFMA) Inc.

extra revenues for new developments that could benefit residents. Rate of Investment (ROI) for garages is higher, causing more useful lands to be occupied by cars instead of low-cost housing. If curb parking and garage parking rate can be up to the market value, extra revenues can be earned. I propose higher revenues from new services and related conveniences uniformly with the same pool of spaces. In addition, removing such spaces to the newly defined suburban parking district (SPD) from CBC will free up the city prime land for other public services.

Establish a Scientific Method for Distributing Parking Spaces with the Right Data Collection
There is no scientific or statistical basis for approving more off-street parking spaces other than empirical formula. Empirical formula so called minimum parking space requirements determined by gross floor area (GFA) of building or occupancy types of Class 1 to Class 9 facilities is used. Classes are defined as Office Complex, Residential Apartment, Health Care Facilities, Industrial or Commercial/Manufacturing Facilities, Religious/Educational Facilities, Entertainment Facilities, Bar/Restaurant, Retail Services and Automobiles Marketing/Repair Facilities.[12] I propose a much more equitable or efficient use of iCloud tools to collect instant data for all kinds of planning of consolidated transportation services including parking and vehicle movement.

Integrated Regulations for On-street and Off-street Parking
Integrating best practices into the current parking paradigm requires multiple stages of planning. I propose new customer friendly regulations to bring all kinds of parking operations under one uniquely defined platform. The first step of integration is developing a common smart-phone or web-based application that serves both on-street and off-street parking. This application would enable drivers to be directed to a vacant space, make a parking reservation, pay for a space, and perform other functions. This application would need to interface with on-street parking (meter space and non-meter space) and off-street parking (includes airport parking, event parking, daily and monthly parking). The second step of integration is to engage and coordinate with key municipal leaders and interested stakeholders. During the economic

---

12 www.houstontx.gov/planning/developRews/docs_pdf/parking

recession of 2008, there was an opportunity to use federal funds in the stimulus package. Unfortunately, the parking industry could not justify allocations of these funds on its own even though the transportation system made a good use of billions for upgrading the Intelligent Transportation Systems (ITS). Coordinated planning among all participants (municipalities, states and private owners) of transportation and parking could have enabled better use of these funds to deliver an integrated parking plan, and enhanced parking security with WiFi networking. Prof. Shoup's book laid the groundwork for fair parking regulations, yet it recommended a slow evolution or incremental change. My book offers a radical departure from an incremental approach to cover many areas not addressed in Prof. Shoup's book. For example, a new approach is required to take the parking industry to the next level of *customer service called e-service* via Robust Web Parking, Truck and Transportation Portal (RWPTTP). It will bring new views of other social, economic and technological changes in transportation services. For example, there are new efforts by USDOT on ITS to advance transportation services through computer aided technologies. ITS enables various users to be better informed and make safer, more coordinated, and 'smarter' use of transport networks with the help of latest information. ITS has also successfully integrated communication technologies and mobility in its infrastructure to reduce environmental impact. There are different interpretations of ITS in the world. In USA, ITS also includes Transportation Service Agency's (TSA's) activities under Homeland Security mainly for the airport and its passenger security. In less-developed countries, ITS is a way to build rural and urban economic development relationships. ITS should include parking as part of its planning for an efficient use of parking spaces, parking security at airports and other vital government installations. The importance of ITS has been recognized by different groups associated with the transportation systems to pursue common technology that could influence our future of transportation services including the environment.

Parking Enforcement

The current enforcement procedure is very crude and labor intensive. It may also be biased because the process lacks automation. There is a general perception that enforcement officers work on a quota to earn a bonus. I propose an ideal environment of self-enforcement policies where ambiguities are removed by electronic data collections. Every event by all stakeholders would be logged and time stamped for proper documentation. Parking rate structures could be based on signed agreements of an account as has been done for the toll services by 'No Stop & Go' surveillance systems or other social media services.

Proprietary Hardware

The incremental upgrade and customization process is very clumsy as much of the hardware is not Internet friendly or cannot be web enabled. It is time consuming because of its dependence on specific vendors. I propose to eliminate all such outdated hardware with software building blocks to emulate hardware functions by software. The process will result in improved performance and simplified online regulations and enforcements.

Arbitrary Rate Increase

Parking is a monopoly business by the city with no fear of competitive pressure. Arbitrary and frequent rate increases with no rationale irritate customers. Here is the best quote from the Internet on the parking rate rationalization.[13]

> Parking spaces are very important to cities. *A city must have enough parking spaces* to provide their residents and their visitors a place to park their car. Since cars are a main factor in transportation, a city must meet the needs of the drivers. If people can't find a place to park, or if they have to pay too much for parking, these people probably won't come back to your city to do some more shopping, dining or spending money in any other way.

Also residents must have enough places to park their car nearby their house and workplace. Unfortunately, parking revenues are not always used for improving parking facilities or customer services. I propose a more equitable solution to contain costs and increase scopes for

---

13 www.priceintelligently.com/blog/bid/182007/6; 'How to Sell Price Increase to Your Customers' by Mark Hunter of Small Business and www.workingflowmax.com/blog/youve-decided-to-raise-rates.

revenues. Coordinated planning of other transportation services may shed new light.

Payment Options

Too many payment collection options cost money to parking operators and owners. Cash, debit cards, credit cards, EMV (a standard and secured payment system designed by Europay, Mastercard and Visa), payment machines, and electronic wallets and each add to parking operators' overheads. Unfortunately, parking operators currently have to cater to all kinds of customers, causing the increased overhead cost of parking. One unique solution for all collections is a pre-payment option that replaces the need for a human interface and thus lowers overhead costs. Virtual currencies such as Bitcoin or Ethereum[14] may also be a contender in this type of system. In addition, auto identification and auto invoicing or even online payments could eliminate such costs.

All of the above ideas scratch the surface only. With an incremental approach, we cannot expect the overarching solution technologies can provide today. Furthermore, a local piecemeal solution would only be a short-sighted stop gap. A national and global strategy must be thought through. For decades, the Internet has played a key role in many new business enterprises. For example, the market value for Internet social media companies that are not even ten years old is over several billions. One can learn from the way the Internet has revolutionized these companies and social lives. My vision is to achieve a similar goal while establishing a new parking platform for the twenty-first century. Key functional enhancements for the proposed overarching solution are:

- Auto identification while parking and then auto invoicing – pay as you use.
- Mobility to activate parking anytime and from anywhere – a nationwide service for uniformity.
- Web-based infrastructure – establishing a future parking Google.
- Partitioning workload responsibilities with end users (like the airline industry).

---

14 Nathaniel Popper, 'Ethereum, a Virtual Currency, Enables Transaction Rival Bitcoin', *NYT*, March 28, 2016.

- Contract based penalties for online services like telecommunication/information technology services.
- Recording all interactions with a time stamp for faster dispute resolution.
- One click solution for the vacation package of hotel, airline, car rental, etc.
- Alleviate the woes Prof. Shoup raised in his two books, especially providing more revenues for freeing up city prime lands.

In summary, the parking solution should be the tail end (last mile) of other integrated virtual transportation services. To mediate among stakeholders one needs to simplify regulations and develop a strong public and private cooperation so that the utilization of these national assets is efficient. That is why I propose that the problem be lifted from the local level to the national level. Such a network of parking resources will enable users to satisfy their needs on demand fairly and equally. Technological maturity is on our side, and can be fully exploited for a revolutionary solution. Parking episodes are almost daily headline news in every local newspaper. Court battles, fist fights with meter maids, bribes, fines, use of cash and unfair regulations taint adverse public perceptions of the parking industry. Sarcastic cartoons, funny videos, and strange comics are very prevalent and reflect the negative public image of many industries. It has grown to be a multi-billion dollar business and the parking industry has its own share. Shouting at the meter maid is a common occurrence, causing a growth of public discontent and adverse images in the parking industry. In other words, there is no perception of civility while treating the customer and the driver. No other service industry can stay in business if a minimum level of civility and courtesy is not maintained. Ticketing and forced collections of escalating fines do not help the industry either. No other service industry uses the criminal court or the judicial system of the country so much as the parking and traffic industry. There must be a friendlier way to charge for services and impose fines for violations and restore civility. Empowering drivers with responsibility to be part of the solution could alleviate some of those public malfeasance and public discontent. Finally, ticketing for one minute violation and or initiating a forced collection of escalating fines is humiliating to the industry. Even under such strict regulations, $20 billion per year of uncollected parking fines remain. Better options for an integrated platform of traffic/parking/toll violations must be innovated. The introduction of virtual parking service offerings or e-service could be one such platform to earn more revenues without forced collections by regulations. Giving up appropriate work responsibilities to the customer is one way to reduce costs as well as earn the customer's respect.

ABBREVIATIONS

Abnb	Airbed and breakfast
AI	artificial intelligence
AITFPA	Affordable Integrated Transportation and Fair Parking Act
API	application programming interface
BDA	big data analytics
BDM	big data mining
BLE	bluetooth low energy
CBC	city business center
DMV	Department of Motor Vehicles
DOT	Department of Transportation
EMV	Euro-pay for Master and Visa Card
EV	electric vehicle
GFA	gross floor area
GHG	greenhouse gas
GPS	geo positioning system
GW	giga watt
HSR	High Speed Rail
IOT	Internet of Things
IPI	International Parking Institute
IPMI	International Parking and Mobility Institute
IT	information technology
ITS	intelligent transportation services
IITS	integrated and intelligent transportation services
JPO	Joint Program Office of ITS and USDOT
LR	Light Rail
MW	mega watt
NPA	National Parking Association
PARCS	Parking Access and Revenue Control System
PARQUEST™	Parking Quest for Route and Availability

ROI	rate of return on investment
RWPTTP	Robust Web Parking, Truck and Transportation Protocol
SPD	suburban parking district
TPD	truck parking district
TPP™	Trusted Parker Program
TSA	Transportation Security Agency
USDOT	United States Department of Transportation
VC	virtual currency such as Bitcoin, Ethereum, etc.
WBPP	web-based parking platform

BIBLIOGRAPHY

Adiv, Aaron and Wanzhi Wang, *On-Street Parking Meter Behavior* (Ann Arbor, MI: University of Michigan Report, January 1987).

Alter, Y. Lloyd, 'The 800 Million Parking Spaces in America have Environmental Impact', December 15, 2010. www.treehugger.com/author/lloyd-alter

Anon, 'Removing the Gate, and Hiring the Right Senior Manager', *Parking Today* (February 2013).

Bigelow, Pete, 'A Big Makeover is Coming to the Parking Garage of the Future – Thanks to Autonomy', *NYT*, July 2016.

Bishop, Rebecca M., and Thomas L. Mesenbourg, *Population and Housing Units Counts*. Washington, DC: U.S. Gov. Census Bureau Report, 2010 Census Briefs.

Buetner, Russ, 'Man Sentenced for Punching Woman in Fight for Parking Space', *NYT*, June 22, 2012.

California City, 'A City Council Report in California While Evaluating WPS Bid', August 27, 2013. www.californiacity-ca.gov/CC/index.php/building/rfp-and-bids

Charlotte, Durham and City of Raleigh, NC, emails from each city manager, 2010–15 citation statistics.

Chatterjee, Amalendu, 'Modernizing Parking Access and Revenue Control System (PARCS) using Parking Web Portal (PWP) for Cost Effectiveness', in *Operation Efficiency, Uniformity and Increased Business Margin*. Exim-soft's internal Business Strategy Report, March 2004.

Chatterjee, Amalendu, 'Enhancing Parking Security Against Terror Threats', *Parking Professional Magazine* (May 2004), 37.

Chatterjee, Amalendu, 'Decriminalization of Parking Services – A Paradigm Shift', *Parking Professional Magazine* (November 2006), 29.

Chatterjee, Amalendu, 'Parking's Impact on Intelligent Transportation Services for Greener Evolution: Part 1', *Parking Professional Magazine* (March 2009), 26.

Chatterjee, Amalendu, 'Parking's Impact on Intelligent Transportation Services for Greener Evolution: Part 2', *Parking Professional Magazine* (March 2009), 40.

Chatterjee, Amalendu, 'Parking Services of the 21st Century and Stimulus Package', Carolina Parking Association (CPA) Conference Presentation, October 2009.

Chatterjee, Amalendu, 'Can Parking Industry Ease Regulations and Bring Civility?', *Parking Today* (December 2012).

Chatterjee, Amalendu, 'Easing Regulations via a Parking Web Portal?', *Parking Today* (February 2013).

Chatterjee, Amalendu, 'A Bigger, Smarter, Simpler and Cheaper – Affordable Parking Act for the Industry', *Parking Today* (January 2014).

Chatterjee, Amalendu, 'The Zero Cost of Parking Payment Processing – Web Account or Bitcoin', *Parking Today* (December 2014).

Chatterjee, Amalendu, 'A National Parking Resources Network? Yes – Here Are the What, Why and How, Part 1', *Parking Today* (March 2015).

Chatterjee, Amalendu, 'A National Parking Resources Network? Yes – Here Are the What, Why and How, Part 2', *Parking Today* (April 2015).

Chatterjee, Amalendu, 'Should Parking Be Part of ITS? If So, Why and How', *Parking Today* (February 2016).

Chatterjee, Amalendu, 'Should Parking Be Part of ITS? If So, Why and How?', *Parking Today* (March 2016).

Chatterjee, Amalendu, 'Parking Lots: Source of Renewable Energy?', *Parking Today* (May 2016).

Chatterjee, Amalendu, 'Parking Lots: Source of Renewable Energy?', *Parking Today* (August 2016).

Chatterjee, Amalendu, 'Why a New Parking Industry Business Model?', *Traffic Infratech* (April–May 2017).

Chatterjee, Amalendu, 'No Meter and No Gate: Auto ID Like Automated Toll Services?', *Parking Today* (October 2018).

Donahuey, Kerry, *State of the Nation's Housing*. Cambridge, MA: Joint Center for Housing Studies of Harvard University, 2017.

Fitzwilliams, Jeff, 'It Probably Happened Just Like This: "Thanks for Those Stinking Tickets, Giuseppe!"', January 2003, www.parkingtoday.com

Harper, Andrew, 'Financial Analysis of the Proposed Third/Second Street Parking Garage', report submitted to Santa Monica City, 2016.

Lockheed Martin, 'Lockheed Martin Parking Catches Sun Power', Press Release, January 26, 2016.

Mazur, Christopher, and Ellen Wilson, *Housing Characteristics: 2010*. Census Briefs. Washington, DC: U.S. Gov. Census Bureau Report, 2010 Census Briefs.

Monahan, Torin, '"War Rooms" of the Street: Surveillance Practices in Transportation Control Centers', *Communications Review* 10(4) (December 2007): 367–389.

NPA, 'Parking Lots and Garages'. NPA internal report. www.npapark.org/08_pubs_04.html.

Robinson, Eugene, 'Let Us Debate the Drones', *News & Observer* Editorial, August 4, 2012.

Scholz, Andreas, 'Parking Technology beyond the Year 2000', www.wpsparking.org.com, (1999).

thestreet.com (2019) 'What is the Average Cost of Solar Panels in the U.S?', www.thestreet.com/technology/average-cost-of.solarpanel

Victoria Transport Policy Institute, 'Transportation Cost and Benefit Analysis II – Parking Costs', (www.vtpi.org).

Wikipedia, 'Intelligent Transportation Systems', September 13, 2012.

Winerip, Michael, 'A Field Trip to a Strange New Place: Second Grade Visits the Parking Garage', *NYT*, February 12, 2012.

CHAPTER 2

Public Perception of the Current Parking Paradigm

INTRODUCTION

I start this chapter with an overview of the word 'perception' because the public attitude to the parking industry is not favorable. It is viewed as frustrating by different publics. Examining the relevance of perception to parking services may shed some light. I have been researching the topic since starting to write this book. A collection of quotes on perception provided some clues to what is needed. These reveal weaknesses and bottlenecks in others' views. This can be a basis for improvement or for radical evolution if so desired because perception points to an underlying truth. People's approval will follow if you provide value and convenience as well as improving the image. That is the connection between future transportation services and these quotations on 'perception' – an abstract view is shown in Figure 2.1. Quotations substantiating the abstract view have been taken from Internet.

A chaotic environment prevails in the parking industry because of the diverse socio-economic backgrounds of people who can afford to buy cars. The noble principle of charging drivers in the busy city is becoming a nightmare to manage – congestion, air pollution, over-regulation and aggressive enforcement. Enforcements, sometimes, are used by cities to raise extra funds for non-transport budgets. Being fair to all stakeholders, whether rich or poor, car owners or public transport users, has become a challenge. As a result, the parking industry has created its own bubble.

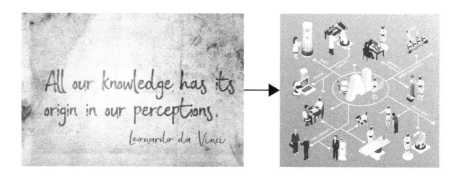

FIGURE 2.1   Power of perception – an abstract view (Shutterstock).

Those of us who are in the bubble need to understand how we are viewed by our neighbors: I mean our customers, drivers, city planners, and business owners. We need to articulate a response to correct perceptions because the parking industry is one of the hottest topics in the news – real or fictitious.

The concept of charging car owners for public parking started almost eighty years back. Slowly, over many years, the system has been manipulated by cities with strict regulations to extract cash from drivers. Eventually, the manual system evolved to a semi-automated one with the sole purpose of collecting money. That is why the base of on-street and off-street parking facilities has grown into a big 'cash cow' industry – almost three or four spaces for one car. Cities cannot be blamed for earning money but what is missing is customer service and fairness on a par with other industries such as telephones, electricity, and the Internet – 'pay as you use'. Parking was under the control of local governments but cities' water, waste collection and transportation services have better customer reputations. Why is the parking service in such a terrible state? Reasonable explanations may be as follows:

- No sense of responsibility (complete monopoly) as long as cash was flowing, with arbitrary rate increase and heavy penalties.
- No competition – very few private operators within a city's or an airport's jurisdictions.
- Strict regulations imposing tickets, fines, and penalties.
- Unfriendly and manual enforcement procedures.

- Underutilization of vast resources in shopping centers and other commercial locations.
- Off-peak and peak rates are not consistent with the market demand.
- Technology was left to the vendor – the result being expensive but sophisticated proprietary hard meters and access devices to meet regulations.
- Web-enabled software upgrades are non-existent.
- Disproportionate overbuilding of parking facilities – on-street parking subsidizing garage parking.
- No consideration of listing parking facilities as national assets with private and public cooperation.
- No linkage of transportation and parking services for a common technology to optimize capital investment and operating costs.

The result of the above is a continuous deterioration of the public image of the parking industry. There is not a single day that goes by without some news report on parking in the local newspaper. This phenomenon has resulted in a huge amusement industry around parking events. I will briefly review cartoons, comedians, reports of fist fights, and funny videos to understand the extent of the damage caused by the attitude of all key participants – city, driver, enforcement officer, and operator. Politics surrounding the city council and the airport authority, including vested interests, are other important factors. Reviews and analyses of these phenomena can provide useful data which are critical if we are to draw lessons on how to improve. A summary is presented below, extracting information without copyright violations from the Internet (Google).

Cartoons

Hundreds of cartoons were reviewed. Many interpretations of those cartoons are possible. Figures 2.2–2.6 show some samples of those. I have grouped them as follows:

- Pure entertainment
- Chaotic or lack of understanding of city irrational regulations
- Lawful and unlawful parking

With his magic power, Merlin had increased parking options.

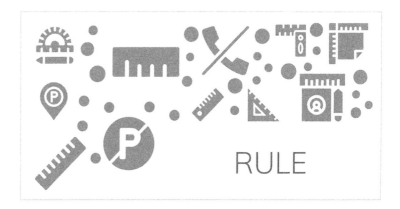

FIGURE 2.2   Cartoon group.

- Lawful and unlawful ticketing to raise money
- Gender chauvinism with manual operation
- Violent behavior including use of guns
- Authoritarian or lack of trust
- Cruising and idling for a spot
- Misuse of handicapped signs

Jokes

Some jokes are out of date in the high-tech arena. Examples are plenty if you do the Google search. Several interpretations are possible. Groupings from these jokes are as follows:

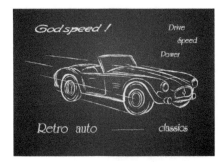

"The best way to start in broadcasting
is to learn how to work the booth."

FIGURE 2.3   Joke group.

FIGURE 2.4   Incident group 1.

- Temporary handmade physical signs: why do you require them in the information age?
- Jokes mostly for entertainment, not practiced while driving or parking

FIGURE 2.5   Incident group 2.

- High-value car gives you auto parking rights – Really?
- All SUVs are compact cars for congested parking lots – Wow?
- Stack parking saves space – laterally or vertically?
- If you want to be on TV, drive away before you get a parking ticket.

Fist Fights/Incidents

There are many incidents of fist fights. They are all real-time incidents. These incidents have increased over time because of growing intolerance and mistrust of officers imposing penalties. Even officers can get involved. Interpretations of these events and incidents are important in considering high-tech solutions. They are grouped as follows:

- Roughing up by cops showing arresting power over ticket issuer.
- Use of guns by drivers to scare ticket writer.
- Other kinds of violence by different parties.
- Lawful and unlawful arrests.

FIGURE 2.6   Incident group 3.

- Liability suits costing municipalities millions.
- Traumatic experience if a child is present.

Funny Videos/Newspaper Articles

Like any other categories, there are plenty of funny videos. Many interpretations are possible. Videos and news articles cannot be shown in the form of texts. Figure 2.7 group 1 illustrates these. They could be interpreted and grouped to highlight for inclusion in our vision as follows:

- Parallel parking – hitting back and forth while parking.

FIGURE 2.7   Newspaper group 1.

- Parking lot accidents – men vs. women statistics.
- Parking skills required to park in specific difficult spot.
- Double parking revenge.
- Illegal parking.
- Towing and high penalties to retrieve cars many miles away from the parking location.
- Inconvenient parking access and exit.

There are millions of those episodes described, creating an industry of its own. In addition, there are books and videos copyrighted for millions of dollars. They have been reviewed in this article to show some embarrassments for the industry. Future embarrassments must be avoided in the modern information age. We must introduce user-friendly customer service, competing with other similar industries, especially when the parking's importance is raised to the national level. Customer service in the parking industry is the worst because of cash payment and unfriendly enforcement procedures accompanied by strict regulations. Why is there so much fuss for the parking? Here are some very simple explanations:

- Human touch and mostly manual operation;
- Attitudes of owners, operators and vendors to change;
- Political manipulation and fear of technology disruptions;
- Lack of automation to isolate elements of conflicts.

In the examples given some fictional elements have been built on facts and real scenarios. Parking owners are, again, subdivided into public and private. In addition, there are parking regulations, parking enforcements, tickets, penalties and parking payments and collections – all of which became the subject to be mocked.

The perception of parking has accumulated over many incidents and over many years. I conclude from the above discussion that there is a lack of civility in the parking industry. Such a lack of civility leads to a worse customer service. Both can be corrected and improved. Social change, availability of technology, inclusion as national assets, environmental considerations, shared responsibilities, customer expectations of mobility, and the example of other industries are the tools to be used to change the parking paradigm to a new platform of civility. The cost of such modernization may be minimal compared to the convenience and other benefits it would bring. I will go on to explore evidence in other chapters to prove this hypothesis, 'the zero cost of civilized parking'.[1]

---

1 IPMI Annual Conference Presentation by Amalendu Chatterjee (Ignite Session), 2014.

CHAPTER 3

Emerging Technologies and Associated Terminologies

INTRODUCTION

In this chapter, the interrelated technologies and terminologies for vehicles, parking, wireless connectivity, worldwide web, and transportation will be described. The parking paradigm cannot evolve without proper consideration of all intertwined areas. The author is drawing on experience as a telecommunications expert for over 35 years, being at the forefront of digital technologies while working with Nortel, GTE and Fujitsu – an era of Internet Protocol (IP), and digital revolution including voice and data integration.

Technology impacts on the parking industry came onto my radar as a serendipity. My friend and I started a company in 1999, Eximsoft International, to develop mobile applications. Our first project was with Ericsson – to develop a secured mobile e-commerce following an early mobile electronic transaction (MeT) standard. We successfully integrated MeT and the Wireless Public Key Infrastructure (WPKI) of Entrust in one of Ericsson's mobile telephone sets for secured transactions – a prelude to the commercial mobile e-commerce of today. The model has been showcased in international conferences as well as in the North Carolina Technology Association (NCTA). The success of the project led us to explore jointly the potential of future mobile applications – the real market killer. During our research, cellphone-based game technology was very hot but we wanted to go beyond the child's paradigm. All team members agreed to investigate mobile parking and transportation services. The last nineteen years of research have led to this book. Exact correlations may not be obvious unless all aspects of the vehicular, parking, and transportation industries are properly reviewed to set the context. The context will hopefully help implementers, and decision/policy makers to study the subject further for its economic and social benefits.

Historically, the implementation of many new but successful technologies is not pleasant one because it disrupts what we have been used to. A paradigm shift in transportation services with parking included over the next century may not be an exception. Similar to other sectors, blue collar workers will likely need to be retrained and re-educated to adjust to new transportation jobs that use software automation. Old jobs may be transferred to countries where cheap or unskilled labor is readily available. According to tech experts, a broad range of technologies reached a tipping point in 2015. They range from laboratory science projects to inventions transforming society and humanity.[1] We need to assess the width and the breadth of such perspectives of new technologies in terms of life span, robustness, and usefulness for creating skilled job opportunities in the parking industry. For example, the telecommunication industry has benefitted from Voice over Internet Protocol (VOIP) to contain long-distance multi-media calls and the airline industry has benefitted with workload sharing with customers via kiosks or home-printed check-ins. If done diligently, a transformation of transportation services will create millions of non-traditional jobs at a rate of several hundred thousand per year directly or indirectly, given transportation services contribute 10% to the national GDP.[2] For example, look at the drone technology that has been making its mark in the war field as well as in consumer commerce. A role for it in parking and traffic management is also on the horizon.

While setting the stage for future, let us not forget the generation gap that exists between the millennial and the boomer[3] generaions as well. Boomers are those born between 1946 and 1964. Millennials (also called Gen-Yers) are

---

1 An article in *Washington Post* by Vivek Wadhwa and reprinted in the *News & Observer*, Science Section, January 11, 2015.
2 USDOT BTS Report, 2004. Kimley-Horn Associates, Inc. 'Parking as an Economic Development Strategy' and Dale Denda, 'Post-Crash Parking Market: Up, Down or Out', *Parking Today*, May, 2007.
3 'Millennials vs. Boomers', *Parade*, August 23, 2015.

those who are born between 1981 and 1997. Boomers hardly saw color TVs, computers, or cellphones when at college. They moved into the dorm with a desk and a bookshelf. Millennials and the following generation would not dream of going to college without access to Twitter, Facebook, Instagram, smart phone, iPad, laptop, and iPOD. The availability of mobile apps and constant mobile connectivity will shape millennials' social behavior.[4] They are accustomed to constant and faster change and growth, with the immense power of graceful adaptability while interacting with friends or with service providers and product suppliers. The Gen-Yers, and those after them, move into the dorm with laptop stand, battery case for iPhone 6 or later models, wireless headphones, smart keyboard, and charging hubs. They are always on the move and downloading mobile apps in their smartphones to get their job done on the fly. Millennial behavior includes online purchases, multiple transactions, mobile fund transfers, and much more – a radical transformation of social norms brought to us by different technological innovations and new capabilities. Can parking be on their radar? If so, what, when, and how? This chapter will address some of those perspectives while cataloguing technologies, connected terminologies, and their relevance for the future of consolidated transportation services. Various researches suggest that Gen-Yers do not want to fool around with meters and gates or cash and swiping cards that slow down their normal lifestyles. They want instant and virtual solutions of the type offered by Uber/Lyft or AirBNB.

This premise should provide enough incentives to explore common technologies to combine the parking and transportation industry, providing economy of scale. Hopefully, information provided in this chapter will bring that argument home. For example, improving the infrastructure of city's congested traffic management via a unique integrated and intelligent transportation service/system (IITS) platform, if well defined, can produce benefits into the next century. The new IITS platform should be an enabling mechanism like social media to cater to this generation Y as potential customer, potential entrepreneur, and also as a future employee. The business model in the transportation industry should suit millennials' future living style – no cash and lots of flexibility to match quick social movement of connectivity, mobile apps, and entrepreneurship.

---

4 Internet of Things (IOT) – Wikipedia. Discussion on Internet, 'Digital Connectedness Good or Bad for People?' initiated by Emerson Csorba and Noa Gafni Slaney, *NYT*, November 28, 2016.

A well-orchestrated list of all advanced, available and relevant technologies consisting of web and related intelligent software that could influence such a paradigm change is required. Specific use of technologies to reconfigure the transportation industry including parking will be highlighted in other chapters. For example, the auto industry could have a future impact with new artificial intelligence (AI) and smart software capabilities. The auto industry is in the process of technological revolution, with transformation in software, hardware, and functions and intelligent screens on the dashboard. Different tax incentives for fuel economy and faster greenhouse gas (GHG) reduction at the Federal and State levels may accelerate the acceptance of such changes. Data collection (speeds, fatalities, performance, loan payment, maintenance, and inspection report, etc.) by auto industries will also influence the parking and traffic industry policy and regulation. Common technologies will provide a needed economy of scale, leading to a national standard with a uniform implementation for customer-friendly transportation services.

There have been many similar examples. Patents and prototypes developed by companies in the past such as Bell Lab, Nortel, and IBM give us an indication of technological benchmarks pursued over many years. Those patents are sold at high prices for new products and services through market demand even though some companies are bankrupt. Prototypes developed that were collecting dust on the lab shelves are getting new life. Voice over Internet Protocol (VOIP), audio-video communication, video phone (modern name – Facetime) and multi-media applications (digital streaming) are some examples. This background information has been given to generate ideas for new business models, process improvements, service scopes for new revenues and improved automation, client identification, and service billing for the combined parking and transportation industries. Next I present my takes on the future outlook for deploying technologies.

Emerging Vehicular Technologies

Vehicular technologies evolved from a simple mechanical device to a sophisticated computer with a screen and many functional buttons. Figure 3.1(a) show a pictorial view of the dashboard contrast. The new dashboard, Figure 3.1(b), looks like a plane cockpit, with automation of multiple functions the traditional vehicle cannot do. These two pictures contrast the dashboard of 1900 (ignition switch and the mechanical steering wheel only) with the modern-day dashboard. Imagine how far we have come in the last 100 years. If this is an indication of progress, the next

FIGURE 3.1(a)   Old vehicle with no dashboard.

FIGURE 3.1(b)   Modern dashboard looking like a plane cockpit.

100 years will be beyond our wildest dreams. Now, imagine the future radical transformation of the intelligent transportation industry. These dashboards may be an indication or incentive for future consolidated road transportation services for generating new revenues. With inputs from parking organizations many additional functional buttons can be added, such as auto piloting – auto parking and auto driving.

The objective of the modern dashboard has been to automate vehicles fully and to connect smart devices and technology gadgets to assist drivers as much as possible, including finding information related to all matters of parking. Some gadgets may be short-lived but the impacts may be long-lasting.[5] Such a transformation could bring about a revolution in intelligent transportation and parking services (ITPS). Hopefully, the latest information about relevant technologies associated with vehicles and their capabilities will provide additional clues for new innovations. Readers will be familiar with different buzzwords and their relevance may suggest enhancements. Enhanced technology in all areas of the auto industry is enabling more efficient, durable, light, stylish, electric, and multi-purpose cars – a computer on wheels. Intelligent software, electronic sensors, light sensors, and built-in camcorders will enable autonomous vehicles to be monitored remotely in the future even if they are parked. Metal parts and their replacements are evolving to become more durable as well as multi-functional, even working as backup batteries. Additionally, new technologies may make traditional GHG and CO_2 tailpipe emissions things of the past. The question is how fast can we let society evolve, given the very high demand for autos in developing and under-developed countries such as China and India? Even in the United States, 17 million fossil fuel new cars are added every year and still greatly surpass the manufacturing capacity of modern electric vehicles. It is given that new technologies may affect vehicle weight, size, and shape impacting future auto parking. Such physical modernization and the resulting software revolution will impact transportation service provisioning, including the design of each parking space lot, for example.[6]

---

5 Farhad Manjoo, 'The Gadget Apocalypse Is Upon Us', NYT, December 7, 2016.
6 Ryan Citron, 'The Future of Smart Parking is Integration with Automated Technology', *Navigant Research*, Jan. 26, 2017.

AUTOMOTIVE HEADLIGHT

The evolution of vehicle headlights has gone through many stages.[7] For example, incandescent bulbs evolved to halogen, xenon, and then LED. Headlamps have definitely gotten brighter but had they become safer? – that was the question insurance companies are asking for the next cycle of headlight design. According to the Insurance Institute of Highway Safety, too much emphasis has been given to the aesthetic design rather than the road performance of lights. After many tests and much research, the Institute has come up with new recommendation for the next cycle of the headlight design or light emitting sensors that must be considered for new automated services that include auto parking.

AUTONOMOUS OR FULLY AUTONOMOUS VEHICLE

Vehicles are being modernized with artificial intelligence (AI) and sensors to be able to be self-driving and self-parking. Initially, manual intervention will likely be required during autonomous driving. Eventually, vehicles will be fully autonomous so drivers do not need to intervene at all. Even autonomous vehicles could override the live driver. Rural interstates may see the fully autonomous vehicles first, because vehicles are confronted with fewer variables and less traffic congestion. Thereafter, they may be introduced in traffic-congested city areas. Such autonomous driving will naturally evolve to autonomous parking, requiring smaller parking spaces with no pedestrian walking space or vehicular maneuvering space. Such automations are equally applicable to passenger vehicles and commercial vehicles transporting goods and people nationally and internationally.

VEHICLE BATTERIES

Both hybrid and electric vehicles need batteries as a source of backup energy. Current batteries are bulky, expensive, and heavy, requiring further research to become light-weight, thin, and flexible. Research ideas include using some body parts of the vehicle as battery sources (anode, cathode, and solid electrolyte in between them). The latest nanomaterials made of extremely thin and strong carbon fiber may replace the vehicle's steel body panels, roof, doors, bonnet, and floor.

---

7 Eric Taub, 'Headlights Get New Attention as More Than a Car Design Flourish', *NYT*, February 16, 2017.

FIGURE 3.2   Future metallic vehicle body with integrated battery.

Such batteries shown in Figure 3.2 can reduce the car's weight by over 15% and can increase the vehicle's range over 100 miles. Thin metallic electrolytes (zinc-air) instead of fluid electrolytes (lithium-ion) between anode and cathode could prevent catching fires as has lately been seen in the Samsung smart phones. Such light-weight vehicles could reduce the load-bearing capacity needed for new high-rise parking facilities.

VEHICLE COMMUNICATIONS – VEHICLE TO VEHICLE (V2V) AND VEHICLE TO INFRASTRUCTURE INTEGRATION (VII)

Vehicle to vehicle (V2V) is a way to share road conditions, traffic jam, and traffic collision information with the vehicle (car or truck) in front or behind you. It enables you to take corrective action when necessary, such as selecting quieter routes via less congested roads. V2V can enhance road safety by warning drivers of potential blind spot collisions. VII, on the other hand, connects vehicles to the Intelligent Transportation Services (ITS) infrastructure for emergency services and applications such as highway traffic management. Highway management services include probing vehicles for traffic, weather, and road surface condition, route advisories, crashes, incident responses, and other warning services.[8] VII could be a useful tool for broadcasting parking space availability to the potential drivers (commercial or private).

MULTI MEDIA INTERFACE OR TOUCH (MMI/T)

MMI/T is a soft touch interface to make controlling functions much easier than dials or voice commands while driving or parking. It enables drivers to input characters on the touchpad to find an address and directions for navigation, enter phone numbers, select songs, and a language for communication. Such capabilities can also be made portable to communicate with smartphones while vehicles are parked.

BIOMETRIC INDICATORS FOR AUTONOMOUS VEHICLES (BIFAV)

Personalization of vehicles may reach another level of sophistication for parking as well as driving. Technology engineers at Continental Automotive Group have been working with cameras for facial recognition, software, and other biometric indicators to automate the driving experience.[9] It is possible to personalize the vehicle so that nobody other than the owner can start the vehicle. Seats can be adjusted to cater to your requirements; the vehicle can play music of your taste and satisfy other wants and desires. This is not just about convenience but also improves the safety and enhances security by monitoring the driver's eye and sounding an alarm if his attention is distracted while driving or parking. BIFAV can prevent the vulnerability of hacking of keyless operation for stealing or breaking in – and is a tool that could be used to configure a trusted parker program (TPP™) or send an alarm if someone is trying to break in. Honda is calling such technology an 'emotion engine', reading driver's emotional state (sadness, happiness, or being angry or drunk). Chemical sensors can be enabled by software to differentiate between a normal driver and a drunk driver to prevent cars from starting electronically.

Faraday Future (an electric company startup) has developed an FF 91 sport utility prototype that detects the car's owner and automatically

---

8 Research reports available in different university (Clemson University, University of Michigan and Berkeley University) and DOT (FHWA-JPO-09 – 038, February 2009) websites.

9 *Biometric Identification Market for ICE, EV, and Autonomous Vehicles, by Authentication Process (Fingerprint, Voice, Iris, Facial, Gesture, Multimodal), Sensor (CMOS, Optical, Retinal), Processing Component, Application, Wearable, and Region – Forecast to 2022*, marketsandmarkets.com, August 2017, Report Code: AT 5499.

unlocks or starts the vehicle. Honda's NeuV, Ford, and Chrysler may also follow the suit. One can hack the system by imitating the image, copying from other online information, so technology has to be perfected by differentiating the live three-dimensional images in contrast to the printed image. Some tradeoffs between convenience and privacy have to be researched with government regulations.

COMPREHENSIVE VEHICLE TRACKING

There are different applications for vehicle tracking. For example, concerned parents could track teenage drivers, or insurance companies could track a vehicle's mileage and speed to determine rates. Additionally, the government could track a vehicle's distance travelled to assess fees, based on miles per year driven on highways. In any case, some of these options may be on a voluntary basis rather than compulsory in a free society. Such tools could also be used for tracking vehicles when parked if you forgot where you parked or lost the ticket in an airport after a long trip.

DRIVERLESS TRUCK AND PARKING

The original ideas for autonomous vehicles were conceived in a Google technology incubator lab. Since its inception, Google has been testing the concept on roads over a decade with a fleet of twelve computer-controlled vehicles. Eighteen-wheelers by Otto Inc. may supersede autos in autonomous driving because of their economies of scale for integrating expensive self-driving technologies.[10] Many states have even passed regulations in favor of autonomous vehicles and trucks. In California, a bill was passed by legislators in 2012 to bring the driverless vehicles onto its roads. Many other states are following suit. There may be initial technological glitches but the systems will be improved and perfected over time as more field experience is gained. Another name for a driverless car or truck is an autonomous vehicle (AV). The idea of full automation in the vehicle industry is slowly becoming a reality, similar to auto piloting of a plane. Far-reaching possible benefits[11] of AVs are:

10 John Markoff, 'Want to Buy Self-Driving Car? Big-Rig May Come First', *NYT*, May 17, 2016.
11 Josef Hargrave, 'Connectivity/Driverless Cars to End Parking Problems', February 14, 2013 available at arup.com/perspectives/driverless cars to end parking problems.

1. Decluttering of city streets

2. Freeing city streets for business and people

3. Driving smoothly and reacting faster than humans

4. Reducing ownership – an upgrade to Uber and Lyft service

5. Providing a new approach for parking, nose-to-tail, side-by-side

6. Reducing time to park in one spot

7. Utilizing unused parking spaces efficiently and saving cities' prime land

8. Delivering higher comfort and convenience for people than they now enjoy with ownership

9. Reducing cost and maintenance headache such as regular oil changes and emission testing

10. Reducing traffic by 40% in inner cities, while people look for parking

11. Reducing parking space requirements by 62%

People do not realize that the surface space occupied by vehicles competes with the space needed for urban housing. They are or will be almost at fifty and fifty occupancy according to most conservative estimates. If we let current empirical city parking regulations stand without modernization, urban development for housing may be difficult. What cities could do with the free space that becomes available when driverless cars operate regularly is beyond the imagination. There could be more space for businesses, and green spaces for recreation among other uses.

Should the parking industry be at par with these modern car manufacturers? Yes, it should. It is advisable for the parking industry to work with these manufacturers to include parking in their grand plan along with smartphone and mobile infrastructure (Ford Media Center Announcement, September 26, 2018, New York, on 'Data Sharing with Uber and Lyft and Cities for Congestion Control'). For example, as soon as a destination is selected by the driver all available parking space information (cost, availability, and distance from the place of visit) should pop up on the driver's dashboard (like Tesla's supercharger locations) so that the driver can select a space at his or her convenience.

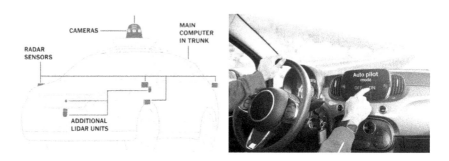

FIGURE 3.3   Autonomous vehicles equipped with lidar and sensors (Uber from Wikimedia Commons Image).

I call it PARKQUEST™ like MAPQUEST. A national database of the parking industry (private and public) will go a long way to building such infrastructure. Also, pre-arrangement of payments can be set up with a national account by the driver to avoid on the spot cash/credit payments delay.[12] That is how one can bring uniformity in the parking paradigm. Self-driving vehicles are equipped with mounted LIDAR units to take 360 degree images, RADAR sensors to measure distances of neighboring obstacles, and a main computer analyzing data to signal to drivers or take autonomous action as shown in Figure 3.3.

DRIVER OVERRIDE SYSTEMS

When cruise control in vehicles was introduced, drivers had the flexibility to override it, especially during bad weather or if tires are not properly inflated. Even in autonomous vehicles, drivers could override many functions to take over the control of self-driving or self-parking. But for *fully automated* vehicles some functions cannot be overridden by drivers, or the vehicles can disregard driver's commands for safety reasons. In the future, for example, vehicles may apply brakes even if drivers have the gas pedal floored. The rapid increase in combination of sensor technologies can shift the priority to full vehicle automation over you to protect lives and surroundings. It may be applicable in the parking scenario as well when vehicles' auto maintenance is triggered before vehicles could be driven.

---

12 Guilbert Gates, 'How Self-Driving Car Works', *NYT*, December 14, 2016.

ELECTRIC VEHICLES (EV)

The rise in manufacture of electrical vehicles is for obvious reasons – because electric vehicles reduce dependency on fossil fuels and create a future cleaner environment. Traditional companies such as Ford, General Motors (GM), Nissan, and many others are working on purely electrical vehicles. Since 2012/2013, production of all electric vehicles has increased, with government tax incentives of different kinds. In spite of such efforts in the United States, demand for vehicles in developing countries like China and India cannot be met. The growth of 17 million traditional cars per year in the United States alone will continue for several decades. All electric cars are an upgrade from hybrid cars that run partially on batteries and partially on gas. Toyota, Honda, and Lexus are pioneers on hybrid cars. Tesla Inc. with electric vehicles (EVs) is the latest auto maker competing in the traditional market with Ford, GM, Lexus, and others. Tesla is producing different models of pure electrical cars which do not use any gas at all. Issues with pure electrical cars are mainly three: batteries to run long distance on a single charge, charging time, and number of charging stations while driving on interstate highways. By 2015, the market was close to 900,000 units worldwide. The market might grow to several billions in the next decade, loaded with features such as auto drive, auto-park, auto brake, auto ignition interlock, and many more, with proper security and safety to prevent accidents, carjacking, and thefts. There will also be more players with innovative technologies, efficient batteries and electrical motors to change the auto paradigm of the next century. In spite of all government subsidies and industry incentives, the forecast may not reach the total size of the traditional market – 2 billion – by 2030.[13] This may be due to the following technological imperfections in spite of low maintenance and running costs:

- Cost effective efficient battery and its replacement frequency

- High initial cost of the electric car development

- Driving range and frequent charging

- Charging time and availability of charging stations

---

13 David Jolly, 'Despite Push for Cleaner Cars, Sheer Numbers Could Work Against Climate Benefits', *NYT*, December 7, 2015.

- Low fossil fuel price and the government's different tax subsidization of oil companies

- Government policies on subsidies and other economic incentives – tax credits

- Regulations on worldwide air pollution and carbon emissions

- Marketing efforts by dealers

All of the electric vehicle models mentioned have battery backups. Several models of electric motorcycles and utility vans are also coming out in the market. Electrical vehicle sales in the United States constitute 40% of the world market and California has the highest concentration, with an infrastructure of highway charging (supercharger stations) and maintenance support.[14] Over the next century, the market volume of electric cars in the United States may reach the same level as today's car market (300 million).

The popularity of all electric vehicles has gone up due to tax incentives allowed by federal government (under American Recovery and Reinvestment Act of 2009 – ARRA) and by some state governments. Other nonmonetary incentives, such as free parking and high occupancy vehicle access lanes, may continue to promote electric cars. Additionally, electric cars aided with solar power will bring a disruption of the oil industry, just as the cellphone did to the landline telephone. One question we should ask ourselves now: what combination of technologies can bring similar disruption to the parking industry and traffic enforcement? There are four obvious changes that should be considered:

1. Every parking space may have an electrical plugin,

2. Real-time one-click event to drive to an available parking space,

3. Integration of parking services with websites such as Priceline. com, Travelcity.com, Hotels.com, etc. for a total holiday package and

4. Online ability by enforcement officers to remotely immobilize cars while parked if warranted.

---

14 Leslies Baroody, Jennifer Allen and Lindsee Tanimoto, 'California Energy Commissions – Tracking Progress, Zero Emission Vehicle Infrastructure', 2016, updated every year.

FLYING CARS (FC)

It may be a fantasy world or science fiction now but may soon be a reality. The Flying Car (FC) has been prototyped by a San Francisco startup, Kitty Hawk. According to an article in *NYT* by Larry Page, a Google founder, the government of Dubai is seriously prototyping the concept.[15] The prototype is an open-seated, 220 pounds contraption with one seat, powered by eight batteries with propellers sounding like a speed boat. Other companies are trying out the technology but with a different approach. One such approach is vertical takeoff and landing (VTOL) planned and designed by Airbus, the French aerospace company, to operate both on the ground and in the air. Of course, there are challenges from the technology and the government regulation point of view, controlling the air traffic by competing vehicles including drones. But developers think such prototypes will generate interest and investments from enthusiasts and hobbyists. Uber, the pioneer of urban transportation services, has also showed tremendous interest and is proto-typing a transportation management service in collaboration with U.S. space agency (NASA) on congested roads[16] – it is called a flying taxi service for people and goods. No doubt, those flying taxis will need to park in convenient open places – another new dimension of the parking paradigm.

PURE SOLAR VEHICLE (PSV)

There is a buzz in the marketplace for the pure solar vehicle. In contrast to electrical vehicles requiring plugins and frequent charging, the energy required to run solar vehicles is obtained mostly from the available sunshine while driving. Like any other vehicle, the solar vehicle is also fitted with a gauge to monitor the range of power available, speed, and odometer. Like electrical vehicles, it may also be backed up by battery power when the sunshine is unavailable. Pure solar vehicle technology is still in its infancy. Designing solar vehicles may be very challenging, because the following factors must be carefully analyzed for the most cost-effective product:

- Positioning or the geometry of the solar panel while driving – horizontal, vertical or otherwise

- Solar panels run only at 15%–20% efficiency

---

15 'Flying Cars Get Uber Boost with NASA Deal', *News & Observer*, May 9, 2018.
16 Kavita Iyer, 'Uber And NASA Collaborate for the Flying Taxi Project', *Techworm*, Nov., 2017.

- Choice of PV cells – either silicon or gallium arsenide

- Availability of the body surface that is exposed to the sunshine

- Maximization of energy extraction for running the engine and the electronic information dashboard

- Target weight of battery requirements to be less than 25 pounds for the backup energy fully distributed in the car real estate

- Maximum miles and the speed that can be driven with backup battery power in the absence of sun

Solar vehicles use very efficient motors and if the solar vehicles are designed perfectly, it will use the energy equivalent of a toaster (2 to 3 horsepower). Such cars can attain speeds over 100 miles per hour – a typical family car. So far, solar vehicles have been used for different races such as World Solar Challenge, American Solar Challenge, South African Solar Challenge, and different University Challenges. Typical speeds recorded in the *Guinness Book of World Records* were over 56 miles per hour in 2014.[17] Without the battery backup, it could reach over 88 miles per hour, eliminating the extra weight of batteries. EVX, an Australian startup located in Melbourne, is planning a commercial solar vehicle called Immortus. Their motto is 'Keep driving as long as the sun keeps shining'. Initially, EVX plans to build 100 cars at a price of over $300,000 – so for wealthy people only. The company is looking for funding but would like to explore niche areas of solar technologies. It aims to perfect solar energy and not to invest in peripherals (like Tesla) of cars already available in the market.

PERSONAL RAPID TRANSPORTATION (PRT)

According to Donn Fichter,[18] PRT is a system of small public auto-mated transport vehicles (designed to carry three to six passengers per vehicle) operating on a specially designed transportation network for cost-effective and faster services. It caters to optimizing an individual's

---

17 Wikipedia. www.solarpondaeration.com and also 'China Reveals New Solar Buses', West Virginia, July 9, 2012. Retrieved Jan. 12, 2013.
18 Donn Fichter, *Individualized Automatic Transit and the City.* Chicago, IL: B. H. Sikes, 1964 and also wikipedien.wikipedia.org/wiki/cabinentaxi.

FIGURE 3.4 Personal transportation vehicle.

transportation needs for low and medium density living areas without owning a vehicle – a modern trend of the millennial. Figure 3.4 shows such a vehicle displayed in an auto show. Several PRT systems have been designed but many have not yet been delivered due to constant changes in technological paradigms. Politico-economic agreements at the local and state government level are important.[19] Key parameters of PRT, also called automated guideway transit (AGT), are optimum location of transit hubs away from large transportation systems such as seaports/airports, railway/subway stations, etc. The hubs, away from population concentration (urban and suburban) centers, also called last mile, are optimally located for efficient frequency of operation (office hours as well as out of hours). The PRT system is similar to other concepts, such as short transportation from the busy city center (BCC) to the so-called suburban parking district (SPD) or the transportation hub.

HEALTH MONITORING

Tracking vital health data of drivers through the steering wheel or seatbelt sensors may be features of future vehicles, as has been trialled by Ford motor company. Vehicles could be paired with their drivers through wearable devices, which could call paramedics if drivers have

19 Richard Gilbert and Anthony Perl, 'Grid-Connected Vehicles as the Core of Future Land-Based Transport Systems', *Energy Policy*, vol. 35, no. 5, pp. 3053–60. doi:10.1016/j.enpol.2006.11.002.

heart attacks or collisions. Drivers, when feeling sick, can trigger full autopilot features of the vehicle and could be taking to the nearest hospital emergency parking space with a signal to the administration.

BLOOD ALCOHOL IGNITION INTERLOCK DEVICES (BAIID)

Ignition Interlock Devices (IID) devices are used to prevent drunk drivers with a blood alcohol (BAC) level above certain thresholds from driving their vehicles. Several states have used IIDs as punishment devices to be installed at the owner's expense, after a first time drunk driving offense. Drivers must breathe into IIDs for the BAC level check before starting their vehicles. The latest infrared spectroscopy technology is being perfected for a better device. Further research is required and work has progressed to automate the system so that the BAIID becomes a mandatory system for all drivers before starting the car for the safety of all – nobody should be able to start the car at any time with his alcohol blood content beyond the regulated limit.

TECHNOLOGY OF THE WORLD CONNECTIVITY AND ALSO PROPOSED RWPTTP

In early 1970, only a few high-speed computers were connected to share research data among research organizations. Data processing was conducted mostly by big hosts with connected terminals. The Internet was limited to a selected few. Today, billions of devices are connected either wirelessly or through wire to link consumers with product manufacturers or service providers. The term Internet of Things (IOT) is commonly used today for interconnections. IOT can be defined as a network of connecting devices with unique identifications, be it machines, computers, objects, animals, or human beings with or without smartphones. The concept seems very noble: autonomous interactions and operations for sharing data without human interventions. The objective is to integrate many day-to-day usable things to provide a business solution. The security challenge is a nightmare while designing such a system because of the integration complexity and the technical immaturity. Any weak security link of a single device could create havoc for other important devices in the network which may carry critical personal and private data. Future applications of IOT and services must take advantage of it for a wider customer base. Figure 3.5 shows the IOT configuration of typical connected devices including doors and

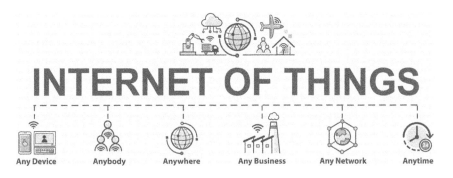

FIGURE 3.5   Internet of Things (IOT) configuration sample.

locks. The concept of IOT can be extrapolated for RWPTTP – connecting all transportation vehicles to the transportation infrastructure for complete automation (be it for human or for goods transportation). Designing a scalable and robust RWPTTP will be a challenge, of course. As in IOT, delivering parking and transportation services to all levels of the community nationally and internationally is possible. Below are brief descriptions of many such devices that may be part of IOT as well as RWPTTP.

Integration with Smartphone

Smartphones can command autonomous vehicles or the vehicle dashboard can communicate with the cellphone and the world when integrated with a vehicle. For example, you can make hands-free phone calls using a Siri-like application to start the vehicle on a cold winter's day, or open the garage. In autonomous mode you could even use a smartphone to call a car from the garage to meet you on a street then drive you to the office. Smartphones integrated with autonomous vehicles and RWPTTP infrastructure may act as on-street meters or virtual gates for all transportation services.

Inductive or Wireless Charging

There are two methods of charging for transferring energy from one source to the other: conductive charging and inductive charging. Conductive charging needs plugs and cables. Inducting charging is also called wireless, plugless or contactless charging. It acts like a transformer in a power station and transfers energy via an alternating electro-magnetic field using inductive coils. The charging station will have a primary coil in the charging plate and the secondary coil at the receiving end as shown in Figure 3.6. They need to be in close proximity to each other. Devices equipped with the secondary coil can run directly or charge a battery with a converter. For example, electric vehicles must be equipped with the secondary coil and a converter to charge the battery either at home or any other charging station like a parking lot. Resonant inductive coupling from one to many can assist these two coils to be apart – the technology has yet to be perfected. There are many pros and cons for such inductive charging. The best advantage is the aesthetic quality of power transfer requiring no cable. The worst disadvantage is that the device cannot be moved around while charging because it has to sit on the charging pad for the whole duration of charging.

Internet Connection

Vehicles with smartphone integration can provide some access to the Internet. But a better option is to have direct Internet access by autonomous vehicles without smartphones. Direct Internet access by vehicles will provide live traffic updates, internet radio, surrounding places of interest, live maps

FIGURE 3.6   Inductive charging principle (Wikipedia).

with cheaper gas prices, laptop connection, and much more. In fact, future vehicles could become WiFi hotspots. Audi sees this as part of the Long Term Evolution (LTE) of autonomous vehicles.

Autonomous Client Identification (ACID)

ACID is a new database with many attributes to identify modern transportation service clients as well as parking service clients or integrated service clients for automated, online, and instant invoice generation. Attributes relate to the account every client sets up prior to receiving the autonomous services. The mechanism for receiving autonomous services is either via the autonomous vehicle or the mobile phone. Any such device becomes the integral part of the future service provisioning. Airlines are in the process of automating and upgrading the boarding and luggage checking process using facial recognition instead of traditional manual processes. Other transportation services may follow suit for the convenience of clients.

Autonomous Invoice for Payment

A natural extension of the autonomous client identification database (ACID) is to design a database of autonomous invoice for payment (AIFP) generation. Databases for such invoice payment generation are: proof of invoice, client reference identification, email, account number, rate and total charge, penalty if not paid by scheduled date, one time or recurring invoices, discount rate if applicable, tax amount, invoice dispute reasons, and any other appropriate localized information. Figure 3.7 shows an invoice sample for an automated NC I-540 toll (No Stop & Go) service, North Carolina.

Personalized Marketing via Vehicle Connectivity

All social media, such as Facebook, Twitter, and Gmail, use ads targeting you based on your buying habits and social behavior. By the next decade when all vehicles will be fully interconnected, different marketers will take advantage of marketing tools to customize their message according to your tastes, location, and driving habits. Hopefully, they will be optional features so that one can use them when needed without driving distraction. The new integrated transportation services platform, RWPTTP, may take advantage of such marketing approaches for new revenue generation and compete with other social media networks for its own market share.

FIGURE 3.7   NCDOT Quick Pass automated invoice sample.

Remote Vehicle Integration (RVI)

If a vehicle is stolen, this feature can catch the culprit. It can also be used when vehicle payments have not been made or too many traffic tickets have not been paid. Such a capability may be an enforcement tool for traffic and parking operations.

Sensors

Sensors are smart devices either integrated with an auto (software emulation) or installed on the ground where autos are parked. Both types of sensors could be either electromagnetic or ultrasonic. Auto-integrated sensors detect the proximity of obstacles surrounding the auto while in motion. For example, your blind spots are protected with signals and you do not have to turn back to reverse the car. Ground sensors are used for parking notifications when a spot is occupied in either on-street or off-street locations. Ground sensors could also be integrated with video cameras to detect a vehicle's license plate when entering or leaving the parking lot.

5G Wireless

Cellular networks have gone through 2G, 3G, and 4G generations. 5G is a new standard for wireless network, providing an access speed of

10 gigabytes per second (gbps). Networks differ by the frequency spectrum, modulation, and bandwidth allocation schemes. 5G is also called the fifth generation network (5G) – an upgrade from the current IP-based fourth generation wireless network (4G) using GFDM, Generalized Frequency Division Multiplexing, as well as Filter Bank Multi-Carrier (FBMC), UFMC, Universal Filtered Multi-Carrier (UFMC) radio technology. Each has its own advantages and limitations and it is possible that adaptive schemes may be employed, utilizing different waveforms adaptively for the 5G mobile systems as the requirements dictate. This provides considerably more flexibility for 5G mobile communications. This access speed is 100 times higher than the current access speed of 1/10th gbps.[20] An article highlighted key benefits of 5G network.[21]

The 5G network will take less than five seconds to download a movie from the current over eight minutes. This will also enable carriers to introduce many new services via smartphone, of course, at a cost.

- The 5G network will also help to connect millions and billions of devices. Examples are: smartwatches, highway sensors, appliance sensors, and many wearable items. The objective is to extend connectivity to rural areas for more revenues.

- Response time of driverless vehicles will be reduced to milliseconds from the current 50 to 80 milliseconds. Such a short response time will eliminate the worries of latency between connected devices such as bicycles for road safety. Ultimately, 5G-based Internet infrastructure will make autonomous vehicles a reality faster.

- The cost of last mile access to Internet will be reduced by 5G. This will enhance smartphone devices to support many new online applications such as parking and transportation including reservations, bus or train arrival, payments, and enhance mobility.

The 5G mobile network could serve as the backbone access network of RWPTTP for all benefits described above during driving or non-driving time.

---

20 William J. Broad, 'Bits: What 5G Will Mean for You', *NYT*, July 16, 2019.
21 Mark Scott, '5G is a New Frontier for Mobile Carriers and Tech Companies', *NYT*, February 21, 2016.

Bluetooth

Bluetooth technology connects wireless devices over a short distance (also called Personal Area Network (PAN) for data or information exchange). It is a fixed wireless standard developed by IEEE as IEEE 150.15.1 but maintained by a Bluetooth special interest group (SIG). It is a wireless alternative to RS-232 data cables to connect devices and operates over short distances using ultra high frequency (UHF) radio waves in ISM band from 2.4 to 2.485 GHz. It was invented by Ericsson (Lund, Sweden) in 1989. The name was derived from the English name (Harald Bluetooth) of a 10th-century Danish King, Blatand/Blattann, who united dissonant Danish tribes into a single kingdom for the spread of Christianity. Originally, the invention was developed to connect wireless headsets but now it has found its way to wider applications such as parking and transportation services.

BLE

BLE is the smarter, more intelligent and low energy version of the traditional Bluetooth operating in the unlicensed frequency band of 2.4 GHz. Original Bluetooth transmits packet using one of seventy-nine channels in the frequency band but BLE uses only forty channels with 2 MHz spacing instead of 1 MHz spacing. Technologically or commercially, it is used to connect all personal devices configured in a personal area network (PAN). It is basically designed and used to enable all machines and devices to be interconnected via the Internet. BLE uses very low energy and can produce over 1 mbps bandwidth for interconnection of many devices.[22] Product and services developed with this technology cover a broad spectrum of applications such as health care, sensing and heart rate monitoring, sports, fitness, entertainment, smart watch, thermometers, and many other applications with potential in other future yet unknown fields. The battery life of BLE may depend on the hardware, range, and the intensity of use, generally ranging from one month to forty months. The parking industry could be one of future application of BLE, as an alternative to the current expensive street meters and garage access with gates. That is where smart vehicles and smartphones are heading.

---

22 BLE description has been posted in Internet, 'WhatIs.com', by Margaret Rouse and contributed by Lisa Phifer and uploaded in November, 2014.

BLE Beacon

You can understand the difference between classical Bluetooth and Bluetooth Low Energy (BLE) from the above description. Once you understand the difference it is easier to appreciate the value of BLE beacons. Bluetooth beacons are hardware devices designed to send quick data between devices. For example, both Android and iPhone operating systems (OS) have been used to enable vehicles with BLE beacons for parking applications. Once you have the device in the vehicle you can transmit your location and appropriate identification via smartphone to a remote location or application. Both in combination act like a Geo Positioning System (GPS) but with a much less impact on battery life and extended precision.[23] Bluetooth beacons work as a one-way device to the receiving device. The receiving device must install an application to use such data broadcast from the beacon. The objective of the installed application is to track users carrying the beacon transmitter. These transmitters are manufactured in a variety of forms – coin size, USB sticks, etc.

Near Field Communications (NFC)

NFC is a set of radio frequency (RF) communication protocols (contactless) between two wireless devices using high frequency (HF) band (13.56 MHZ) within the proximity of 4 cm (or 1.6 in) following ISO standards – one may be a mobile phone such as a smartphone and the other may be stationary device such as a desktop with wireless capabilities. NFC provides a low-speed peer to peer connection setup to bootstrap more wireless applications such as contactless or mobile payment, electronic ticketing, social networking, sharing contact, photos, videos, and files. NFC is basically a subset within the family of RFID technology. The main difference is that RFID communications need a reader, a tag, and an antenna, and can cover up to 100 meters within the same 13.56 MHZ HF band. NFC devices can work as both a reader and a tag. General RFID readers send signals to the tag via the antenna for specific data and the tag responds accordingly. RFID tags could be active with their own power source for a broadcast capability. Passive tags with no power source can communicate up to distance of 25 meters in three different frequency ranges – Low Frequency (LF)

---

23 Wikipedia on Bluetooth Beacon.

125–34 KHZ, High Frequency (HF) 13.56, and Ultra High Frequency (HFC) 856 MHZ to 960 MHZ. NFC has become a standard and unique feature of the latest smartphone so that simple tasks such as contact information or photographs can be shared between two smartphones by tapping devices together. RFID and tags are widely used to 'Stop and Go' control of parking space entry.

Radio Frequency (RF) and Beacon

RF Beacon is a transceiver device (mostly transmitter) usually located in geostationary and inclined satellites. Such devices receive or transmit continuous or periodic radio signals containing limited data on a specified radio frequency. The purpose of these electric or electromagnetic beacons is to broadcast their fixed location so that direction-finding systems (radio operated) on ships, aircrafts, and vehicles can determine the bearing to the beacon. The transmitting function relate to telemetry data and meteorological information. These beacons are also named with specific applications such as space and satellite radio beacons, driftnetbuoy radio beacon, distress/emergency beacons, and much more.[24] In addition of air and sea navigation, RF beacon has applications in propagation research, robotic mapping, radio frequency identification, and indoor guidance, as with real-time locating systems (RTLS) like Syledis. In fact, any AM, VHF, or UHF radio transmitter can be used as a beacon for a direction-finding system. The principle is applicable to infrared and sonar beacons as well. Limited use of these devices has occurred in parking services and more use is expected in the integrated RWPTTP infrastructure.

WiFi

WiFi stands for Wireless Fidelity. It is an IEEE 802.11 (a, b or g) standard for short wireless communications using either 2.4 GHz frequency spectrum (802.11b and 802.11g) or 5 GHz frequency spectrum (802.11a) to derive different data speeds at different distances. 802.11b and 802.11g products are compatible with each other but not to 802.11a products. WiFi compatibility is very important for the WiFi alliance group so that all products from different vendors work together. The group has specified rigorous testing requirements for certification. 'WiFi Hotspot' is a very popular term in the WiFi network. A hotspot is basically an access point (AP) of wireless LAN

---

24 www.spaceacademy.net.au/spacelink/satbcns.

through which one or more wireless devices are connected to the Internet. WiFi can also be used for peer to peer communication between two or more independent devices without AP (Figure 3.8). Some hotspots provide free services and some charge you for the connection. Many hotels and cafes provide a free WiFi connections for your cellphone or computer as part of their main service package. Unfortunately, nobody guarantees the security (free from hacking) of such free packaged services. Autonomous vehicle connectivity as well as smartphone connectivity to parking/ITS infrastructure via WiFi (especially in the garage environment) is key for introducing automation transportation services from all locations.

Geo Position Systems (GPS) or Global Positioning Systems

GPS is a satellite-based navigation system made up of a network consisting of twenty-eight different satellites owned and operated by the US Department of Defense (DOD). There are similar systems deployed in Russia as well as in Europe. Even China and India operate their own local GPS. GPSs pinpoint velocity (if moving), time and location for tracking a receiver anywhere on earth using at least four simultaneous satellite calculations. It has three networking components: satellite in the orbit, user with a receiver, and a control system on earth. Since early inception of the concept in 1960, the system has found diverse use. Examples include: locating a vehicle your children are driving, tracking movements of different animals or birds in the forest, locating shipwrecks in the sea, and other civil or military applications. Imagine a national database of all parking spaces that can be accessed via a vehicle or smartphone, so that a driver knows available spaces to park for his/her convenience. Sensors on the ground are not necessarily required. Vehicles with DLE beacon,

Peer-to-Peer / Ad-Hoc

FIGURE 3.8   Peer to peer communications directly or via an access point.

smartphone, and GPS capabilities can be used to determine total counts of all parking spaces – free or occupied.

Global Navigation Satellite Systems (GNSS)

GNSS is a globally recognized generic and standard term used for global navigation covering land, air, and sea using satellites. GNSS provides autonomous geo-spatial positioning with multiple earth orbit satellites in multiple different orbital planes. Michael Venezia gave a vivid description with global variations of the term in Wikipedia. GNSS has different regional variations for naming. For example, in USA, it is called global position system or GPS, and in Russia, it is called GLONASS. Other operating names are Galileo, Beidou, etc. The purpose of multiple satellites in multiple orbits is to improve accuracy, redundancy, and availability if line of sight is obstructed. Satellite systems rarely fail –if one fails, GNSS receivers can pick up signals from other available systems in the orbit. Applications of GNSS have moved from earlier use in the military to industrial tracking/mapping of devices, machinery movement, sea vessels, air navigation, and autonomous vehicles. Wider applications are expected to emerge in the parking and transportation industries soon. For example, RWPTTP may be one such new idea to monitor all open and occupied parking spaces on a real-time basis broadcast to all intended users. These users are either using their own vehicle or a public transportation system for the last mile.

CONNECTED INFRASTRUCTURE

For the rural as well as global economy, connectivity by sea, rail, planes, or roads is as important as connectivity by wires, satellite, wireless, fiber, or copper. The former is used to transport goods and people and the latter is used to transport information for business exchange and information exchange to be part of variety of decisions. In this chapter, two such infrastructures' roles are described at all levels of such information exchange. They are telecommunications infrastructure and the Internet infrastructure. Different service platforms are then built upon such infrastructures to provide end user services and other applications. A telecommunications infrastructure is the basic level of physical end to end connections. An Internet infrastructure consisting of network and higher layer connections based on the open system interface (OSI) is built upon the telecommunications infrastructure to provide a variety of network services or applications to satisfy consumers' needs as they arise. Internet infrastructure cannot be built without the prior presence of telecommunications infrastructure.

Telecommunications Infrastructure

Alexander Graham Bell invented two-way analog telephone conversations in 1876 based on analogue theory multiplexing voice channels in a physical path with dedicated access (most expensive component) and shared backbone of switching. That is how we got the current worldwide telephone infrastructure using copper loops at the access, a switching system, and long-distance cables, undersea cables, 4G wireless (or latest 5G), and satellites for the backbone of the telecommunications network. A dedicated communications channel and required bandwidths during conversations end to end is derived through the switching system. Dedicated bandwidths during conversations result in high costs of telephone conversations. Then came the digital technology with pulse code modulation (PCM) of 64 kilo bit per second (kbps) for voice communications, resulting in multiple voice channels over one physical wire. Even then, data, telegram, and voice used different dedicated channels within the same telecommunications infrastructure resulting in no cost benefits. With the invention of packet data (store and forward techniques), Internet protocol (IP) and voice over IP (VOIP) came the revolution in voice communications and the death of telegrams (long-distance messaging). Real-time voice and data can now be integrated sharing digital channels, which was not possible decades before. Direct shared access to the end user using wireless and satellite and eliminating last mile of copper or fiber reduced the cost further. Such cost reduction was possible in both the access and the backbone by a new switching technique called packet switching. Digital video and data streaming took the digital communications to the next level. WhatsApp, Webchat, Skype, and Vonage took full advantage of such disruptions to lower the telecommunications infrastructure cost end to end. Sharing the same channel bandwidth requires software building blocks and high-speed processing power of computers. In other words, it is an effort to build the virtual Internet infrastructure for all point to point and broadcast communications. The bottleneck of such virtual infrastructure was the last mile of access to the end user. It was always expensive because of dedicated resources. With the advent of wireless, WiFi, and other shared cell towers, it has been possible to take telecommunications to the remotest rural areas of underdeveloped countries. Results include: educational enhancement, information gathering and sharing, health improvement, new business enterprises, economic growth and globalization of human development at a speed that could not have been thought of a decade before.[25] Because of such disruptions, billions of people are using the resources of the end-to-end infrastructure, lowering the cost of use and leading to additional innovations by common people. It is perceived that the end device of this robust infrastructure, the so-called smartphone, could replace local banking, or other day-to-day online business transactions, nationally and internationally. Such virtual service offerings have increased the mobility of people via smartphone applications. Such robust telecommunication infrastructure helped many Internet related virtual business platforms – Google, Facebook, Amazon, Cloud Computing, and the like. The latest is the virtual Blockchain (or Bitcoin) technology for payments. Hopefully, the parking industry with careful planning can take advantage of all available technologies to disrupt the current paradigm to create a virtual platform to offer many new services and applications. Key to such success is the creation of a national online parking database, correlation of the parking database with the vehicle occupancy anywhere, automated billing to the owner of the vehicle and other related information for a friendlier and cost-effective service such as has never been seen before.

Internet Infrastructure

Internet infrastructure, though built exploiting telephone network resources, is different than the traditional telephone or television network. It is built on the basis of packet switching concept using TCP/IP protocol for connecting computers and users for information sharing. ARPANET, a research collaboration network, is the pioneer of the current Internet and kept evolving with the pace of computer speed and real-time business needs. The extent of Internet is limited to the reach of the telephone network. In general, the network infrastructure definition is very much dependent on the context it is used for. For example, in the context of ITS and parking services, it may be defined

---

25 Gaia Pianigiani, 'Internet Throws Lifeline to Family Business in Small Town in Italy's South', *NYT*, December 8, 2016.

as: 'Installations that form the basis for any operation or system to offer services to clients with user friendly interfaces', interconnected by variety of hardware and software components. Internet resources with their infrastructure are also exploited by cloud computing companies to connect computing resources and end users. Distributed computing with Blockchain payment processing may be the next phase of evolution of the Internet platform. For societal living and the rural/urban economy to function, the infrastructure is characterized as, 'Fundamental facilities such as roads, bridges, tunnels, water supply, sewers, electric grids, telecommunications and systems serving a county, city, or area, including the services and facilities necessary for its everyday functions'.[26] To implement RWPTTP, we will need a network infrastructure at different levels, starting from the fundamental layer of telecommunications consisting of access and backbone made of wireless, fiber, satellite, or copper. After the fundamental facility interconnection at a physical layer comes network infrastructure layer (switching layer) and then comes the Internet or service platform and higher application layer connections. After the first three layers of infrastructure are chosen, a hardware and software platform will be defined for writing specific applications and interfaces for services to the end user. Figure 3.9 show pictorial explanations of key relationships when security is superimposed to prevent hacking at each layer of the infrastructure for improved big data flow and layered service provisioning capability Figure 3.9(a) highlights three layers of outsourced elements and Figure 3.9(b) additional breakdowns of security layer elements from the implementation and deployment point of view. In general, the Internet is a digital platform designed for carrying voice, video, data, and multi-media data catered to a specific application. RWPTTP may be one such candidate to use all these media for future transportation services. Additional definitions of Internet related elements for such a visionary deployment may be helpful.

Webmail

Webmail, also called web-based mail, is a modern application of sending mail and documents electronically. It competes with traditional mail services such as United States Postal Service (USPS), UPS, and FedEx.

---

26 Jeffrey Fulmer, 'What in the World is Infrastructure?'. *PEI Infrastructure Investor* (July/August 2009): 30–2.

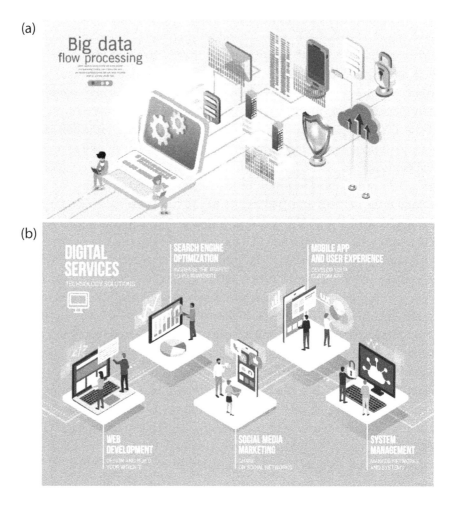

FIGURE 3.9 Infrastructure – a sourced service protected from hacking (a) Secured big data flow infrastructure and (b): stock-vector web development secured communication technology.

USPS's annual revenues have fallen several billions of dollars due to the wider popularity of webmail. Webmail is enabled with the help of the Internet, mail servers, a webmail client computer, a unique electronic address for individuals, and web browsers. Each electronic address is formulated with a certain combination of a service provider name, purpose of the organization (profit or non-profit, government, educational institute), and country in some cases. Companies that have successfully implemented webmails are AOL (AOL mail), Google (Gmail), Yahoo (Yahoo mail), Microsoft (Outlook), and others. The evolution of webmail is very dramatic starting from simple text to the modern day multi-media mails for commercial applications. The initiator of such services started with a concept such as Internet Service Provider (ISP) – content service provider that is different from the telecommunications service provider (infrastructure provider). There are many issues of security, privacy, hacking, and spams associated with these electronic exchanges. Sophisticated encryption technologies and security algorithms (Norton, McAfee, and many more) are being deployed to prevent such misuse of resources online. Many software packages are also updated online to secure them from the latest threats. Webmail is a disrupting technology for USPS, almost making it bankrupt. One wonders if there similar emerging technologies outlined in this chapter will disrupt current parking and transportation services. Perhaps such technologies could integrate future intelligent transportation services for the benefit of all stakeholders?

Web Browser

This is a software tool used to navigate in the Internet and is well defined in Wikipedia. Each individual web page, image, and video is identified by a distinct Uniform Resource Locator (URL), enabling browsers to retrieve and display them on the user's screen. A web browser should not be considered to be the same thing as search engine because the latter is a website such as google.com that stores searchable data about other websites. But the user needs to have his/her own browser to customize its own request. Popular browsers that are being used are: Chrome, Firefox, Safari, Internet Explorer and Edge. The RWPTTP website must conform to these browsers once deployed.

The objective of the browser navigation is to retrieve, present, and traverse software information resources residing in the form of uniform resources identifier/locator (URI/URL) or emails. The browser also helps to access other information resources following different web hyperlinks. These resources may be in the form of audio, video, images, or other forms of multi-media content. This soft tool (browser) is designed with software programs. Since its inception in 1990,[27] the

---

27 'Tim Berners-Lee: WorldWideWeb, the First Web Client'. W3.org. Retrieved December 7, 2011.

web browser has gone through many forms of sophisticated iterations. Lately, mobile browsers have been added. Out of all those listed, mobile browsers for smartphones and Google Chrome for desktops are gaining popularity for their rich functions and features.[28] Web browsers consist of a user interface, layout engine, rendering engine, JavaScript interpreter, UI backend, networking component, and data persistence component. These components achieve different functionalities and together provide all capabilities of a web browser.[29] There may be further enhancements required in this area for new traffic and parking applications using smartphones.

Worldwide Web (WWW)

WWW or simply the Web is a collection of applications stored in databases or in servers for visual or interactive exchanges via the Internet. The concept was introduced by Shannon in 2010[30] but was initially developed by Tim Berners-Lee in 1991. Related concepts required for a fully functional web are: domain name server (DNS), URL, browser, hypertext transfer protocol (http), hypertext markup language (html), etc., so-called internet standard protocols for web browsing as described below. The power of WWW can never be underestimated. All social media have used WWW to create a brand of businesses. Examples of powerful WWWs are Google, Facebook, Amazon, eBay, and many more. They each targeted a core market, and created a niche customer base. Let us narrate some additional elements of WWW for a better perspective on RWPTTP's future potential web configuration.

Domain Name Server (DNS)

A domain name is well defined in Wikipedia as an addressing and identification mechanism of devices and applications such as a computer, website, etc. belonging to an organization with string of words (for quick memorization) instead of string of numbers. The Domain Name System (DNS) is a standard administered by domain name registrars and developed by Internet Engineering Task Force (IETF). DNS is

---

28 Tali Garsiel, 'Behind the Scenes of Modern Web Browsers'. Retrieved October 12, 2013 from taligarsiel.com/projects/howbrowserswork1.htm.

29 Ibid.

30 R. Shannon, 'The History of the Net'. http://yourhtmlsource.com/starthere/historyoftheinternet.html#arpanetandonwards.

organized in a hierarchy with a nameless root. After the root, the most common known domains are: com, info, edu, net, org, country, etc. It is mostly used for the identification of network resources following IP protocols so that proper addressing of each resource is possible in a machine-coded computer network such as the Internet. Second or third level domain names in the hierarchy are reserved for end users who wish to connect local area networks to the Internet. In general, Internet Protocol (IP) resources are: personal computer, application, websites, or any other service etc. used to access the Internet. As domain name is a unique identity of an organization, it is OK to choose a domain name corresponding to the organization's name, helping users to access them easily. Generally, Uniform Resource Locator (URL) in the Internet browser is a tool to show the appearance of a host component in the address bar while accessing the website. As per statistical data collected in 2017, 330.6 million domain names had been registered[31] and more are added every day. RWPTTP may be the next one if implemented.

Uniform Resource Locator (URL)

URL also known as Universal Resource Identifier (URI) is used in reference to web pages in the address bar of a browser.[32] Interchangeably, they are used to locate information resources on a computer network, and are always used in the context of web pages and hypertext transfer protocol (HTTP), such as www.rwpttp.com if RWPTPP becomes such a computer resource (a host) for integrated transportation services. To refer to a file name or an application, the URL address could be as: http://rwpttp.com/index.html. Here, index.html indicates a directory and a file reference.

Hypertext Transfer Protocol (HTTP)

HTTP is a standard set of rules and regulations for transferring texts, graphic images, sound, video, and other multi-media files. Such rules and regulations act as a request/response protocol in a client–server computing environment. For example, you use Internet browser in your laptop (a client) using a URL to retrieve an application in a

---

31 Wikipedia.
32 Tim Berners-Lee, 'Uniform Resource Locators (URL): A Syntax for the Expression of Access Information of Objects on the Network', World Wide Web Consortium, 1994.

google search engine host (acting as a server).[33] The server having all related resources such as HTML files, and other contents returns a response message to the client. A web browser is an example of user agent (UA) software whose specifications have also been standardized by IETF. Other types of user agent are web crawlers, voice browsers, mobile apps, and other software that access, consume, or display web content.

Hypertext Markup Language (HTML)

HTML is a markup computer program for creating web pages and web applications. It was conceived first by Tim Berners-Lee, a physicist and researcher, and then adopted by IETF for standardization. It is widely used by web browsers to interpret and compose text, images, and other contents into visual or audible formats in a page. With Cascading Style Sheets (CSS) and Javascripts, it forms a triad of cornerstone technologies for the World Wide Web[34] (WWW).

Power of Worldwide Web (WWW)

The history of the worldwide web is not very old. Many successful companies have popped up in the last decade. Their growth rate is higher than any other known traditional industrial growth. Google, Facebook, eBay, Amazon and many more have market values over several billions of dollars, creating a brand of product and service offerings out of this Internet infrastructure nationally and internationally in a very short period. Some of them may be characterized as search engines but many of them are encroaching on each other's territory to try and capture market share. The success of these websites is determined by the volume of subscribers – specific to a country or globally. Their market valuation also indicates their potential to change the society. Table 3.1 gives an example of leading market penetration in China and globally.[35] Each of these companies has developed a marketing strategy to grow. They also have a business model for increased revenues while accommodating new technologies with evolving platforms for global as well as regional markets. Many entrepreneurs of products

33 Roy T. Fielding, James Gettys, Jeffrey C. Mogul, Henrik Frystyk Nielson, Larry Masinter, Paul J. Leach, and Tim Berners-Lee, 'Hypertext Transfer Protocol', IETF's RFC 2616, June 1999.

34 HTML Living Standard (http://html.spec.whatwg.org/multipage/) ISO/IEC 15445.

35 Justin Kerby, 'Here is How Much Facebook, Snapchat, and Other Major Social Networks are Worth', *Social Media Today*, May 2016.

TABLE 3.1 Subscription Volume and Market Valuation of Many Social Media Company Built on Internet Infrastructure

Social Media Company	Market Valuation ($B)	Global Market Share by %	China Only Market Share by %	Global Share by % without China
Facebook	434	65	30	80
Google	739	52	39	60
You Tube	90	45		59
Twitter	13	41		50
Linkedin	26	23		30
Instagram	100	18		25
Qzone	1	21	75	
Youku		18	55	
Sin Weibo		30	80	
Tencent Weibo			70	
Renren			50	
Tudou			45	
Kaixin			37	
51.com			25	
Pinterest	11			20
My Space	3			18
Tumbler				14
Orkust	0.150			16

and services are on the same bandwagon. It is questionable if it is a 'real thing' or just a fad, as analyzed by Jayson DeMers.[36] Each social media company has created a specific marketing brand over the years, taking advantage of the human behavior change. It is a plus for many entrepreneurs to associate with these brands and create their own brands through such associations. Brands have many market advantages as they

- Keep up with technology updates
- Learn about the latest innovations of product and services

---

36 Jayson DeMers, 'I Demystify SEO and Online Marketing for Business Owners', *Subscriber*, August 2014.

- Participate in promotions

- Provide useful feedback to cater to a particular need

- Interact with diverse groups for ideas and changes

- Make a quick decision on products/services

- Are confident and prioritize correcting defects and errors

These media companies do not have any products other than the advertisement platforms they have created via Internet. It is an irony that these companies earn steady income from advertisements from other entrepreneurs. The other source of income is via an exclusive right to subscribers' information and selling information to those who need it. The legality of such an approach may be questionable in legal terms. Here, one can see the opportunity of the RWPTTP platform to market its transportation services with its own and unique brand as well as an advertisement media for other related products and services. Associated benefits of such approach are many: exploiting the latest technologies, meeting the latest market trend with smartphone applications in the age of social media, developing loyal customers and brand loyalty, becoming part and parcel of the everyday life of users, having active participation by customers and vendors, expanding the base of Internet subscribers, globalization and regionalization of market segments, decreasing ongoing costs for all stakeholders, improving customer experiences and confidence, etc.

The above dynamic forced the Internet through many stages of evolution. Such evolutions or transformations due to modern technology have disrupted many traditional businesses and established many social and business paradigms. The modern Internet stands as a real-time and online platform developed with robust software and hardware Sustainability Infrastructure with interconnecting devices.

The Parking Portal (RWPTTP) integrated with a transportation portal could also be a one such autonomous platform in the age of autonomous vehicles with a market value of billions. Currently there are two major platform characterizations. They are: Exchange platform and Maker platform. There are subcategories under each of those major categories. All social media brands have been mapped under two major categories and many subcategories. They are shown in Table 3.2. Visit http://technoratimedia.com/wp-content/uploads/2013/02/tm2013dir.pdf for information. Alex Moazed in his Internet blog

TABLE 3.2 Categories of Social Media Platform For Application Development Choice

Exchange Platform	Makers Platform
Service Market Place	Closed Development Platform
• airbnb	• Tridium
• Hotelstonight	• Fitbit
• uber	• salesforce
Product Market Place	Controlled Platform
• Amazon	• Apple
• eBay	• Windows8
• Etsy	• Google sky service
Payment Platform	Open Development Platform
• Paypal	• Google sky services
• Square	• Linux
• livedup	Content Platform
Social network platform	• Twitter
• Facebook	• Instagram
• Linkedin	• YouTube
• nextdoor	
Communications Platform	
• Skype	
• Whatsup	
• wechat	
Investment Platform	
• CircleUp	
• Prosper	
• lendingClub	

writing has given additional clarification as he defines end-to-end business solutions consisting of enabler/enabling platform and enabled platform for future applications. Such background information could be useful when designing the RWPTTP application. The artificial intelligence (AI) and cloud computing technology will impact these platforms and future applications. For example, cloud computing and BDA is creating a buzzword in the marketplace for new brands and new applications. RWPTTP may have the advantage and an opportunity to be more creative to compete better with current social media companies for consolidated transportation.

In this context, it is important to have additional backgrounds about the Worldwide Web or WWW. The Worldwide Web or WWW is an umbrella used by almost all industries to promote business and customer reach electronically under Internet infrastructure. It is similar to real estate bill boards along interstate highways for advertisement. The difference is that it works on electronic highways. Under this WWW umbrella, an industry-specific URL is designated for identification such as www.google.com or www.fb.com. The consolidated transportation and parking industry URL such as www.rwpttp.com could be created to include Robust Web Portal for Transportation, Truck, and Parking (RWPTTP). RWPTTP documents in the website will be formatted in a markup language called HTML (Hyper Text Markup Language) that supports links to other documents, as well as graphics, audio, and video files.[37] It started with 'Web 1.0' in late 1990 with limited contents and Internet bandwidth. It then quickly evolved to 'Web 2.0'[38] when the Internet infrastructure expanded in storage capacity and bandwidth with many sophisticated tools and APPs such as[39] blogs, micro blogs, podcasts, social media, Wikipedia, etc. There may be further evolution to Web 3.0 or so. Other Web related definitions are also enumerated for reference, with a possible use while implementing new service domains in the integrated transportation industry.

WEB STORE

Everybody is familiar with the English word 'store', the place you go for shopping products and devices. These physical stores are capable of selling products and receiving payments for products either in the form of cash or credit/debit cards. Modern websites can create a virtual store with similar functions. Virtual stores are capable of selling products as well as services in the form of applications usually called apps. They receive payments online and offer an alternative to physically going to the store and carrying the stuff after the purchase. Web stores deliver products to your doorstep or directly to your cellphone or computer. Such e-stores fall within the jurisdiction of generic term

---

37 Webopedia definition.
38 Terry Flew, 'Differences between Web 1.0 and Web 2.0', *New Media* (a technical journal), 3rd edn, 2008, and also in Wikipedia.
39 Introduction to Web 2.0 Technologies by Joshua Stern, Ph.D., and available on www.wlac.edu/online/documents/web2.0v.02.pdf, 2017 and www.youtube.com/watch?v.

called e-commerce. E-commerce may entail some or all of online goods and services (e-books, streaming media), retail services (banking, DVDs by mail, food and flower delivery, grocery, pharmacy, and travel), marketplace services (advertising, auction, comparison shopping, social commerce, community trading), mobile commerce (payment, and ticketing), customer services (call center, help desk and live software support), and electronic purchases.[40] There may be a new scope for parking service offerings using e-commerce.

WEBSTAGRAM, WEB4. AND WEB SUDOKU

These are additional terms used in the modern social media depending on the country and special interests. One can google each of the terms for additional explanation. The intention is here to simply set the reference for the appropriate use in the context of the transportation industry, if required. Human interactions have been replaced with automation to connect buyers and sellers for the convenience and instant gratification of our daily lives. We are living in the digital world. In this world, you become an object, instead of a human being, with an identification number and a password. Take for example the modern banking service. Your salary is paid online, You can automate your payments online, If you get a check, you can deposit it in the bank with the help of smart mobile phone applications, You have a credit/debit card to spend, you can see the statement online and analyze your spending habits instantaneously, by day or month. Not only do you know about your affairs, but the whole world knows about your habits and movements. Cloud computing is another platform derived from the Internet platform to move desk-top computing to central, shared computing. Because of innovative reservation platforms like Priceline, Orbitz, and others, you can make all reservations for plane tickets, hotels, and cars with one click. Hopefully, consolidated transportation services may one day reach that stage of being a one-click solution.

Platforms have created a revolution. Very soon all platforms will be able to be supported through smartphones.[41] The parking and transportation industry is in a nascent state. Exploring a combination of platforms could carve out the future for mobile parking applications.

---

40 Wikipedia.
41 Alex Moazed, 'What is Platform', January 2014, Google Search.

POSSIBLE ENABLING TECHNOLOGIES FOR PARKING AND TRANSPORTATION SAFETY (RWPTTP)

Light Detection and Ranging (LIDAR)

LIDAR is a device to detect the size, the shape, and the distance of an object with the help of ultraviolet, visible, or infrared light and reflection. LIDAR sometimes is called laser scanning and 3D scanning, with terrestrial, airborne, and mobile applications according to Wikipedia. Data collected by LIDAR paired with a camera can be analyzed by computers to make decisions in the field. LIDAR can target a wide range of materials, including non-metallic objects, rocks, rain, chemical compounds, aerosols, clouds, and even single molecules. A narrow laser-beam can map physical features with very high resolution; for example, an aircraft can map terrain at 30cm (12 in) resolution or better. LIDAR has been used extensively for atmospheric research and meteorology. It is being used, in conjunction with artificial intelligence (AI), to design autonomous vehicles – self-driving and self-parking to avoid collisions, accidents, and other safety concerns. For such safety functions, LIDAR is fitted on the top of the vehicle to have 360 degree views of neighboring objects while driving or while parking to remind the driver of the vehicle's exact location as you leave the car park.

NEW LOOK OF AUTO DASHBOARD – SOFT CLICK ACTIVATED FUNCTIONS

The modern vehicle dashboard is becoming a driving, information, and other communication tool. The cockpit-like functionalities help troubleshoot if the car does not move. It is like a laptop computer without a keyboard but many functional buttons in front of you enable you to drive from point A to point B. While driving, it also helps you to be connected with the rest of the world without distracting you from the road safety. You will be able to activate and deactivate auto driving and parking, similar to today's cruise control on the highway.

The dashboard is part of auto interior and exterior design, which is highly regulated for the safety of occupants and to prevent objects from causing collisions. The dashboard is an engineering marvel filled with technology, art, and architectural design.[42] It is very complex but on a par with other

---

42 John Roe et al., 'The Definitive Guide to the Modern Automotive Interior: Best at a Price, Deep Dives, and More', *Car and Driver*, March 2015.

handheld devices. Modern dashboards are attractive to the new owner, hiding all ugly secrets of wires, motors, plumbing, airbags, seatbelt clinching, etc. underneath. An appealing but beautifully designed final dashboard puts important components in the right place for easy use. Of course, the design will evolve as the car function evolves so that important functions are at one's fingertips while driving. Mostly, soft buttons will dictate the following functions in future to make it a complete automated vehicle:

- Auto/bio ignition without key
- Motion activated door opening/closing with proper biosecurity features in place
- Command activation not to fall asleep
- Do not disturb warning so that drivers are not distracted
- Auto pilot to activate auto driving or parking
- Automatic navigation to a free parking space close to the destination
- Speaker on to tell you every move the auto is making
- Road and weather condition information
- Hands-free calling
- Important reminders (i.e. taking prescription drugs on time)
- Secured bio ignition interlock to prevent drunk driving
- Trip purpose and travel information
- Hours of parking and fee calculations after parking while exiting the parking lot
- Data collection and analysis for security, safety, and new policy development
- Warning if speed limit is crossed

LIVE WINDOW DISPLAYS OR HEAD-UP DISPLAYS (HUDS)

This HUDs technology converts the windshield or the glass into a transparent touch screen. It may be the same information you get from the dashboard or the smartphone or new information available by Internet connection. Temperature, tire gauges, and other mechanical functions of vehicles can also be displayed in larger and clear scripts right in front of the driver rather than at other spots around the car, a phone, or the radio. This would help a driver better focus and reduce accidents. HUDs may even avoid potential legality issues regarding the use of phones, GPS, and other Internet devices while driving. In fact, HUDs bring all these diverse gadgets and screens into one front screen, allowing drivers to read, text, and email, or even find their relative positions without removing their gaze from the road. Once vehicle manufacturers make this offering as an integral part of vehicles, it will revolutionize autonomous driving, transportation, and parking, providing all necessary conveniences to drivers, operators, and policy administrators.

ASSOCIATED TERMINOLOGIES TO CONFIGURE RWPTTP

Parking Data

The International Parking Institute of USA (IPI), now called International Parking and Mobility Institute (IPMI), has initiated a consortium in collaboration with various European parking associations. This is the first step towards developing a parking standard internationally – a common language of terminology definitions for sharing data across industry jurisdictions. The consortium agreed to define data elements for public and private parking owners, operators, and service providers. The objective is to use a standard format to facilitate the exchange of appropriate information among themselves as well as with the outside world. In their first report,[43] they identified only three data elements for a possible standard requirement. They are: Place, Rate, and Occupancy as Phase 1 standard. For Phase 2, they identified another three data elements: Transaction, Enforcement, and Event Data. This is a very good news for the proposed RWPTTP implementation when the final agreement on an international standard is reached. With such standards, it is possible to go beyond parking services – the essence behind RWPTTP visionary thought of integrating transportation services with mobility when autonomous vehicles will dominate our society in the

---

43 Alliance for Parking Data Standards, *APDS Overview,* V1.0, October 2018.

next decade. It will also enhance private and public cooperation for integrating total transportation services of the next decade. More work will be required to satisfy diverse requirements – the requirements of object oriented format agreements rather than the content itself, because the content will vary location to location as well as the company to company – but common terminology is critical to make sense of the content.

BIG DATA ANALYTICS (BDA) OR BIG DATA MINING (BDM)

Big data for any corporation means the volume of data (structured or unstructured) it stores from different connected and mobile devices or network sources online or offline but mostly via automated collection procedures.[44] The volume of data may not be that important but what you do or how you use these data (in other words to convert them to useful knowledge) in day-to-day business is very important. The term 'Analytics or Mining' is used to process this volume of data to add different values, such as to discover patterns, unknown correlations, market trends, client preferences, and other related information to the benefit of service providers, marketers, and other corporations who are involved to make business, strategic, operational, and research decisions on many issues. Imagine the volume of data generated and captured by a billion devices connected to Internet globally, including streaming data of different forms, with many variables in the cloud computing infrastructure deployed by different companies. The cloud computing infrastructure has also developed many tools to harvest market and business values of these data, catering to a specific corporation's needs. Added values may include faster development of products and services, lowering costs of data research, flexible management, customer satisfaction, and potential future application developments not even imagined presently. Big-data adoption efforts and tools could unearth many intelligent and valuable factors for many other organizations. Parking and intelligent transportation services (ITS) may fall under such category. BDA or BDM is the process (much faster and eloquent than the

44 Mathew Burt, Mathew Cuddy, and Michael Razo, *Big Data Implication for Transportation Operation, An Exploration.* FHWA-JPO-14–157. Washington, DC: US Department of Transportation and Intelligent Transportation Systems Joint Program Office, 2014; Mike Robertson, 'Is Big Data Coming to a Parking Facility Near You', IPI Report, 2017 (www.parking.org). Nicole Ybarra, 'Work Smarter Not Harder – Leveraging Big Data', *Parking Professional Magazine*, May 2018.

traditional database management tools) to generate actionable business insights from early raw data collections to integrate parking and ITS. This could help to build a nationwide Robust Web Parking, Trucking Transportation Portal (RWPTTP) – the basic ingredient for 21st-century urban/suburban transportation systems combining many elements of modern technologies is mobility. For example, autonomous and connected vehicles, mobile customer devices, autonomous parking with no meter or no attendant with gates, autonomous transportation services, autonomous traffic monitoring by drones, autonomous client identification, and autonomous payment will dictate a new virtual transportation paradigm. Big data analytics is the best way to synthesize its usefulness (data capture and management – DCM) by exploring travel pattern – business or pleasure, online and instant congestion control, customer behavior, traffic safety enhancement, reliability improvement, local parking and transportation hubs to avoid city traffic and carbon emissions, parking frequency, last mile mode of transportation (Uber/Lyft or own vehicle – ride sharing), customer preference, occupancy, operational efficiency, revenue maximization, cost reduction, and much more.

BLOCKCHAIN PAYMENT

Blockchain is an underlying digital foundation that supports many applications and Bitcoin may be a subset of Blockchain for financial transactions or other vetted documents sharing. But the objective of Blockchain is to reach out far beyond Bitcoin or Ethereum, a competition with Bitcoin comprising of blocks that hold batches of time stamped and encrypted transactions. It is so named because the technology is comprised of blocks that hold batches of time stamped and encrypted distributed transactions.[45] IBM has a slogan, 'Blockchain can do for business what Internet did for communication'. How? Businesses of all kinds exchange values (goods, services, money, and data) every second of the day and we are moving to online and secured transactions. According to many experts, Blockchain executes many more of such transactions in a much faster, efficient and better way.

Bitcoin is the early version of Blockchain implementation to explore the impact of digital currency launched in 2009. Bitcoin also called

---

45 Wikipedia. Manav Gupta, 'Blockchain for Dummies', 2nd IBM Limited Edition, August, 2018.

cryptocurrency is a decentralized digital or electronic payment system. It is an innovative person-to-person (P2P) cryptocurrency of the Internet invented by a group of programmers. In many ways, Bitcoin is similar to any other conventional network like Visa or Paypal, though, it differs in two important ways. (1) Nobody owns it, managing the process to make profits. Bitcoin is a pure peer to peer structure with hundreds of computers (called miners) over the Internet working together to process transactions. (2) It has its own unit called Bitcoin for monetary transactions. Its main purpose is to avoid U.S. central banking regulations including exorbitant fees for national and international money transfers. Though both the Fed and investors are considering a safe harbor for virtual currency. Bitcoin's currency value has taken off since its inception, from a fraction of a dollar (US$0.30 in 2011) to many hundred dollars (US$600 in 2016), drawing the attention of the venture capitalists. Bitcoin enthusiasts think it is an innovative platform for a new generation of financial services. Bitcoin acts as token between two individuals or corporation national or international which can be exchanged for major currencies such as the US dollar. 'The Bitcoin symbol is clearly popular enough that it should be in Unicode', said Ken Shirriff, the author of the proposal, in an interview with *Bitcoin Magazine*. 'Getting the Bitcoin symbol into the Unicode standard was clearly the right thing for Unicode and the right thing for the Bitcoin community'. Bitcoin or similar virtual currencies can reduce the transaction costs of any transportation and parking services in future – a focus for RWPTTP for advance or automated payments. Virtual currencies have many advantages for money transactions over the traditional banking system: less cost, better safety and security, and more efficiency.[46]

EMV

EMV stands for European Master or Visa card and has become the latest standard for smart credit and debit card payments. EMV is based on a chip or IC technology embedded in the card. User data is encrypted in the IC card in addition of the magnetic stripes for backward compatibility. With the invention of the virtual coins we can group

---

46 Amalendu Chatterjee, 'The Zero Cost of Parking Payment Processing – Web Account or Bit Coin', *Parking Today*, vol. 19, no. 12, December 2014.

future payments for products and services into three categories: (1) traditional methods such as credit/debit card, cash, check, (2) e-wallet such as Google Wallet (MasterCard), Apple Wallet, V.me (Visa), and (3) virtual currency such as Digital Wallet Plus (DW+) and Bitcoin (Digital Mining). It is very expensive for vendors to support multiple platforms. Eventually, they all may converge to a single payment method that parking vendors and operators may support for cost optimization and driver's convenience. Bitcoin for parking payments may be that optimum choice.[47] Bitcoin may be one unique option in the payment world that may be self-protected and self-secured without a sophisticated password mechanism.

INFORMATION TECHNOLOGY (IT)

Humans have been storing, retrieving, manipulating, and communicating information since 3000 BC when writing skills were developed. The term 'information technology' in its modern sense first appeared in a 1958 article published in the *Harvard Business Review* by authors Harold J. Leavitt and Thomas L. Whistler. Modern IT is so diverse that the exact definition may be perceived in different ways: academic perspective, commercial/professional perspective, and application perspective. In an academic context, the Association for Computing Machinery (ACM) of IEEE defines IT as a program (undergraduate and graduate degree) that prepares students to meet the computer technology needs of business, government, healthcare, schools, and other kinds of organizations.[48] Graduate programs are involved for additional future evolution of the field. In the professional and business context, the Information Technology Association of America (ITAA) defines IT companies as a group as the 'tech sector' or 'tech industry'. It further clarifies that tech sectors are involved in the 'the study, design, development, application, implementation, support or management of computer-based information systems'. In the application context, the responsibilities of those working in the field include network administration, software development and installation, and the planning and management of an organization's technology life cycle, by which hardware and software are maintained, upgraded, and replaced. The

---

47 Ibid.
48 Wikipedia.

corporate value of information technology lies in the automation of business processes, provision of information for decision making, connecting businesses with their customers, and the provision of productivity tools to increase efficiency. In the context of applications, people in charge of IT assume responsibility for selecting hardware and software products appropriate for an organization, integrating those products with organizational needs and infrastructure, and installing, customizing, and maintaining those applications for the organization's computer users. Computers are connected to store, retrieve, transmit, and manipulate data, or information often in the context of a business automation and cost improvement covering techniques for processing, the application of statistical and mathematical methods to decision making, and the simulation of higher order thinking through computer programs. The IT operation for the RWPTTP functionalities will need a robust and scalable platform to be discussed later.

ENCRYPTION

The best definition of encryption is (Wikipedia):

> Encryption is the most effective way to achieve data security. To read an encrypted file, you must have access to a secret key or password that enables you to decrypt it. Unencrypted data is called plain text; encrypted data is referred to as cipher text.
>
> (Wikipedia)

Apple has been using encrypted software in iPhones. Lately, WhatsApp has started using encryption technology in their messages. Encryption has become a hot topic for privacy, national security, and terrorism (such as the San Bernardino, CA, shooting). The terrorist or the shooter in this case used encryption in their cellphone and the government could not easily decrypt the cell phone to connect this shooter to other terrorist groups to prevent similar events in future. The government requested Apple Inc. to decode this specific shooter's phone but the company refused to comply. The case has gone to the Supreme Court. It has since been resolved in favour of the government with the help of a third party.

PARKING STANDARD

For a long time, players in the parking industry did not pay much attention to standards and uniformity. There were three reasons for this: local government monopoly of city parking, vendor proprietary products, and no private and public cooperation for cost control and social justice. In addition, local governments introduced their own enforcement policies to increase their revenue shares. The result was customer inconvenience and many complicated enforcement policies. Now simpler regulatory policies are required nationwide because of technology and market pressures. APDS by IPMI is a good start but more is needed. APDS initiated a standard on data definitions only. More areas need to be addressed. Examples are: product interconnectivity, simplified enforcement policy nationwide, regulatory policy relating to urban housing development, formation of a national database, software emulation of hardware, declaring parking spaces as national resources, use of these resources for renewable energy, parking data partitioning of RWPTTP for revenue shares (like the telephone network), centralized management and administrative center, etc.

SOFT METER APPLICATION (SMA™)

SMA™ is software application developed by Eximsoft International while working with Ericsson to replace on-street parking meters with a cellphone. In fact, the application can also work in a garage environment without a gate. Prerequisites of this application are that all parking spaces be marked with unique IDs such as number, street, block, city, and state. All drivers are also required to register with the website of the application to validate his/her account. In exchange, the driver gets a parking permit number in his cellphone. Once the driver occupies a spot s/he calls an 800 number to activate the parking permit with a duration for debiting his/her account. After the expiry of this the driver has the option to extend the time or to leave the spot. Once the spot has been left the account is not charged any more. If the driver forgets to deactivate the permit the system can automatically deactivate it if another driver occupies the spot. Smartphone technologies have the potential to reach a whole new level while creating a more user-friendly environment for meterless parking. SMA™ may be seen as a preamble to future RWPTTP because of its visionary ideas of replacing meters with smartphones and reducing costs by software emulations of hardware functions.

TRUSTED PARKER PROGRAM (TPP™)

Eximsoft International owns the TPP™ trademark. It may also be known as Trusted Traveler Program (TTP™). The idea was conceived after 9/11 (terror attack) to prevent an Oklahoma type garage bombing. In other words, the program, if implemented, can secure all parking spaces around critical government buildings, nuclear facilities, all airports, sea ports and railway stations. The theory is to collect information of on drivers and vehicles that enter the parking lot, like TSA's 'global entry'. For example, it may avoid the 300 feet zone restriction creating a traffic congestion at airports' entry during high alert by allowing only trusted drivers and vehicles. The process will include the following:

- A registration process where passengers and vehicle profiles are collected for automated entry

- A PIN code is provided during registration for parking or entry authentication later

- Integration of wireless public key infrastructure (WPKI) for mobility along with authentication/authorization for activation

- Easy location of cars when tickets are lost and exact position is not remembered

PARKING/TRANSPORTATION KIOSK

A kiosk, in general, is an electronic device or a computer with either a touch screen or a keyboard located in some convenient or high-traffic location as a customer interface to the specific application. The purpose of kiosks is to provide information, render services, or rent or sell goods to people. Most of the modern kiosks will have Internet connections. Examples of typical kiosks are: bank ATMs for cash, electric utility for payment, information systems, airport traveler check-in, hotel lobby check-out, shopping mall for direction, hospital lobby for patients, parking ticket payment, etc. An interactive kiosk is a specialized computer terminal with specific hardware and software combinations for functionalities providing communication, commerce, entertainment, or education. The kiosk design for RWPTTP operation will have similar functionalities in addition to specific transportation functions.

DEVICE EMULATION

According to the online dictionary, emulation is 'when one system performs in exactly the same way as another, though perhaps not at the same speed'. A typical example would be emulation of one computer by (a program running on) another. You might use an emulation as a replacement for a system whereas you would use a simulation if you just wanted to analyze it and make predictions about it. There are two types of emulation – hardware emulation by software and software emulation of hardware. Hardware emulation by software is mostly used to verify complex electronic circuit design for its functional performance. Software emulation can be explained in two contexts. In the first context, one computer system or the host program behaves like another computer system called a guest.[49] In the second context, one hardware device function is completely replaced by a software program or cheaper hardware device with a different configuration. For example, the modern email system with the help of computers, Internet connectivity, and a mail server is emulating the traditional mailing system via the US postal service. Similarly, one can think of replacing the street meter and the garage gate with cheaper or simplified substitute devices such as smartphone, geo positional system (GPS), and the computer programs connected by Internet and application servers. Emulation is generally perfected with the help of repeated simulation with variable data of the real environment and prototyping functional characteristics by computer-run programs. This concept of hardware device emulation by software will play a key role when we will speak of new paradigms and new disruption technologies. Such technologies will transform the traffic and parking industries to an automated system of the 21st century, replacing outdated street meters and gates usually called the Parking Access and Revenue Control System (PARCS).

APPLICATION PROGRAM INTERFACE

API is an intelligent way to enable hardware or software to connect the current abilities of different devices and programs to new devices and programs. API is written in a particular computer programming language with certain rules called protocols, libraries, routines, tools, and

---

49 Wikipedia, September 17, 2015.

portability interfaces to enhance the compatibility and portability of different software. The biggest benefit APIs provide is to customize vendor-independent applications and services to reduce costs. For example, Android ON (X) enables you to extend the capability of the Android phone using a free API license to remotely program it. This gives users an extension of unlimited customized applications based on individual rules and constraints. Parking applications (open space location, its route, and rate) could easily be embedded in the Android phone. For iPhones; however, an open API is not available and a paid development license would be required to customize applications.

ABBREVIATIONS

ACID	autonomous client identification
ACM	Association of Computing Machinery
AGV	automated guideway vehicle
AIFP	autonomous invoice for payment
AM	amplitude modulation
AP	access point
APDS	Alliance of Parking Data Standards
APP	application
AV	autonomous vehicle
BAIID	Blood Alcohol Ignition and Interlock Device
BCC	busy city center
BDA	big data analytics
BDM	big data mining
BIFAV	biometric indicator for autonomous vehicle
CO2	carbon dioxide
COMTO	Conference of Minority Transportation Officials
DCM	data capture and management
DNS	domain name server
~~**EMV**~~	~~European Mastercard & Visa~~
EV	electric vehicle
FBMC	filter bank multi-carrier
FC	flying car
GBPS	giga bytes per second
GFDM	generalized frequency multi-carrier

GFDM	generalized frequency division multiplexing
GHG	Green House Gas
GDP	gross domestic productivity
GNSS	global navigation satellite system
GPS	geo positioning system or global positioning system
HTML	hyper text markup language
HTTP	hyper text transfer protocol
HUD	heads up displays
IEEE	Institute of Electric and Electronic Engineers
IETF	internet engineering task force
IP	internet protocol
ISO	International Standards Organization
ISP	internet service provider
IT	information technology
ISM	industrial, scientific and medical
ITAA	Information Technology Association of America
LAN	local area network
LPD	license plate detection
LIDAR	light detector and ranging
LTE	long term evolution
MET	mobile electronic transactions
MMI/T	multi-media interface touch
NCTA	North Carolina Technology Association
NFC	near field communication
NYT	*New York Times*
OCR	optical character recognition
OS	operating system
OSI	open system interface
PARCS	parking access and revenue control system
PCM	pulse code modulation
PAN	personal area network
PRT	personal rapid transport
PSV	pure solar vehicle
RF	radio frequency
RFID	radio frequency identification
RTLS	real time locating system
RVI	remote vehicle integration
RWPTTP	robust web parking, truck and transportation protocol
SIG	special interest group

SMA™	Soft Meter Application
SPD	suburban parking district
TPD	truck parking district
TPP™	trusted parker program
UFMC	universal filtered multi-carrier
UHF	ultra high frequency
URL	universal resource locator
USB	universal serial bus
USPS	United State Postal Service
VOIP	voice over IP
V2V	vehicle to vehicle
VII	vehicle to infrastructure integration
WAN	wide area network
WPKI	wireless public infrastructure
WWW	world wide web

CHAPTER 4

RWPTTP Technical Design Outlines

INTRODUCTION

The main reason for formulating the RWPTTP platform is to reduce the construction of additional parking spaces. The purpose is to increase the efficiency of utilization, including look and feel of parking services nationwide. Why? There are more than three parking spaces per vehicle. More parking is being constructed simply because of existing regulations and a lack of vision. For example, for every new vehicle sold there is a plan to build another three spaces. Of course, there may not be enough space where parking is needed the most. What is the end result of constructing more spaces? Plants and human beings cannot breathe comfortably. Land is being lost for urban development – housing, business, and recreation. We need to relocate and reorient the physical distribution of these spaces. In addition, we need to integrate transportation services with parking. For example, there is not enough truck parking space on the interstate highway corridors. That is why we need a new parking paradigm with a new infrastructure. Technology, with its associated tools and innovations, can implement for such a vision, which will bring cooperation among all stakeholders, including private and public owners.

There has been a dramatic change in the vehicle design which also influences parking. All vehicles are electric, digital and interconnected (shown in Figure 4.1). All vehicles are capable of collecting data as they

FIGURE 4.1   Connectivity world of the future transportation services.

travel (shown in Figure 4.2). People are becoming mobile with smartphone applications as well. All these phenomena are part of new transportation automation. Such new phenomena help us think of a new parking paradigm. For example, Intelligent Transportation Services (ITS) have come a long way. Do you remember who invented the Internet? It was a government initiated product in 1960. The challenge at that time was to share research data among more than three high-speed computers. Today, the Internet has transformed our daily social and economic life by connecting any device (cellphone, smartphone, iPad, iPod, laptop, street parking meter, garage access gate) and by supporting any application (social media, match making, online education, mobile e-commerce, and much more) at our fingertips. Similarly, ITS is a government-initiated project to enhance surface/road transportation by utilizing the best modern technology. Road/surface transportation means the efficient and timely movement of people and goods by car, bus, truck, or railways from point A to point B under normal conditions or in a national or regional emergency. ITS is a federally sponsored project with yearly funding provided by the U.S. Department of Transportation's Office of the Assistance Secretary for Research and Technology (OST-R). (See more at: www.

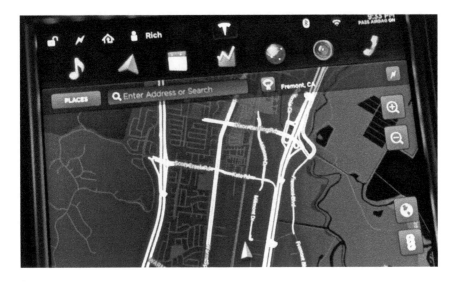

FIGURE 4.2   Screen icons of Tesla Model X for the scope of data generation.

its.dot.gov/faqs.htm#sthash.jx28sSxq.dpuf.) This research and investigation work started in early 1990 and has gone through many years of funding in addition of ongoing implementation by the U.S. Department of Transportation (USDOT) and the corresponding State Department of Transportation (State-DOT). The latest vision has been documented in the ITS 2015–2019 Strategic Plan by USDOT and the office of the Assistant Secretary for Research and Technology, ITS Joint Program Office. Key elements of the vision have been outlined as:

- Improvement of transportation safety and mobility

- Efficient and effective interstate traffic management for toll services and ticketing

- Enhancements of traffic decongestion for American productivity

- Minimize environmental impact (GHG and CO2) – renewable energy, fuel efficient cars, electric cars

- Integration of advanced communications technologies for mobility and flexibility

- A robust and modern transportation infrastructure (drones may be part of the equation)

- Deployment of a broad range of wireless and wireline technologies

- Decision making based on online, real-time and instant information

- Best use of car connectivity (driverless driving and parking) embedded in artificial intelligence (AI) with the transportation infrastructure

The strategic objective is to build an ITS framework (web based and virtual) focusing on road vehicles with new capabilities such as:

(1) Automation to transfer some amount of vehicle control from the driver to the vehicle,

(2) Future Ttansportation systems with emerging services such as light rails,

(3) Enterprise data collection capture from sensors, mobile devices, and connected vehicles for future research activities,

(4) Interoperability of effective devices and systems,

(5) Accelerating deployment as and when required and

(6) Enforcement for real-time and online resolution in collaboration with all stakeholders.

Should parking follow ITS's example? Many parking professionals see the potential for economic, professional, and social growth in linking parking to ITS. IPMI has formed an ITS task force to explore the possibilities and opportunities of such interconnections. The ITS task force is seeking professional input to raise parking services to the next level of sophistication. In fact, many benefits may be derived from the right approach, including federal funding.

What would the parking vision be like?

- Declare parking resources as national resources to initiate public/private cooperation for the best use of these resources (especially prime city land).

- Simplify regulations – three to one space to car ratio is not acceptable.

- Explore common technologies for both ITS and parking visions – a way to reduce the cost of technology development.

- Use emerging technologies for virtual parking, mobility and automation.

- Decriminalize parking services, without tickets and escalating fines, for social justice.

- Formulate a national standard for sharing the national parking database for online demand like airline seats.

- Establish a national robust parking infrastructure with smart/intelligent technology connecting to ITS infrastructure for information sharing.

- Have vendor independent hardware connectivity with API for instant and new parking applications/services.

- Define a suburban parking district (SPD) as the transportation hub for city travel.

- Provision truck parking services along US highway corridors for rural economic development, defining a Truck Parking District (TPD).

- Coordinate TPD with smaller autonomous trucks to deliver door to door goods and services (last mile).

- Stop big trucks running on highways during commuting hours; instead use smaller autonomous vehicles for door to door delivery (last mile).

- Keep up with the ITS pace – software services, pay-as-you use, and workload sharing (changing reservation, printing boarding pass, and check-in).

It takes a long time to upgrade from one vendor to a different vendor. Most of the hardware is not Internet friendly with application program interface (API). Here are some of the new tools for the 21st-century parking industry instead of meters and PARCS.

- ○ Software emulation of hardware devices – cost reduction. Software building blocks are keys to emulate hardware. Bluetooth, automatic vehicle identification (AVI), high-speed cameras, drones, and other devices with auto connectivity can accelerate the meter and gate replacement. Centralized software and database will also make end to end operations efficient. An Internet-based business model helps customization of services with a national account.

○ Use of Internet for mobile applications – keeping pace with the social media. Generation Y takes full control of smartphone mobile apps such as banking, purchase, and travel. A parking app is yet to make that list. Websites such as Priceline.com, carrentals.com, and hotels.com help them to meet their all travel needs. To complete the end to end travel plan, we need a parking website nationwide, or parking could be included as an icon in any of the above websites.

○ Real-time online infrastructure – going virtual with Virtual Parking Web Portal (VPWP). A national database with availability to identify all parking space owners and their locations. This is required for revenue allocations once online reservations using websites are made, similar to airline, hotel, or car rental reservations. VPWP is a concept published in the *Parking Today* magazine, December 2012, which upgrades the traditional parking services with software building blocks, database, and computer networks connectivity of mobile devices including smartphones.

○ Deployment of convenient parking district (PD) to link city parking with transportation – a prelude to ITS and introducing a dual track system with light rail. A district is a type of administrative jurisdiction or territory managed and dedicated by each municipality for the purpose of parking only. Drivers will be encouraged to park in these designated areas. Each parking space of these designated areas will be in the database for online review by all. PDs may be located in suburban areas close to the city covering east, west, north and south boundaries. Once parked, people will be transported to different parts of the city by bus or light rail. Buses will run as frequently as required, with stops on demand. Many auto related businesses such as car wash, tire change, gas station, and repair or even dealerships can grow in these areas – decongesting the current city infrastructure. Auto drive and auto park capabilities of newly designed cars will further refine the PD definition.

Overall benefits to be derived if we are successful in our articulation of bringing parking within the ITS building blocks are:

- Managed traffic with fewer vehicles on the road looking for parking
- Effective truck operation with end to end view for national pride (well coordinated with commuter traffic and weighing stations)
- Common platform to compete with private transportation companies such as LyftLine, UberPool, General Motors, and others
- Interaction among all agencies during disasters
- Workload sharing like airline check-ins by all stakeholders
- Software building blocks for updates and new services
- Single payment for many transportation and parking services
- Toll by plate payment with no 'Stop and Go'
- Auto enforcements to avoid biased or unfair ticketing
- Online configuration and reconfigurations of ITS for expanded parking applications

Parking professionals can see the potential economic, professional, and social growth in linking parking to ITS. There must be a cheaper solution – going virtual with software and Internet for the next century. The current parking industry is much too fragmented. Each operator or owner has their own system with no national or international standards for uniformity. Vendors have been taking advantage of this, resulting in expensive solutions with minimal or no flexibility for easy upgrades. Vendors have vested interests not to make the system web-enabled and inter-connectable for mobility, making changes to new services either impossible or time consuming. Such an approach worked well in the last century but we need a new approach with well-equipped automation for the next century. Solving parking woes will not only bring benefits to the transportation industry but parking must be brought up to standard first and then the total transportation solution can be addressed with a common platform. That is what I will outline here.

The current parking configuration has a huge embedded infrastructure with poor user experience and convenience. Market forces dictated incremental addition of technologies, complicating the embedded base with more cost. Examples include mobile payment and advanced reservation within the installed infrastructure. We need a complete overhaul with new infrastructure to combine parking and transportation services. Let us review some progress that is already on the way before a final strategy can be recommended.

The first step of design objectives is to consolidate all parking scenarios under one software configuration irrespective of the current hardware configuration and their locations. The parking industry has not yet joined the mainstream of all other national transportation services such as toll, traffic management and intelligent transportation services automation. Traditionally, revenues are collected by some sort of enforcement mechanisms based on rate and duration policies. For on-street parking, meters are installed or time zones are established for collecting revenues. For off-street parking, gates with access cards (credit cards, spitted tickets, monthly pass, etc.) or attendants are placed to collect revenues. Over the years, meters and gates have become sophisticated but not fully automated or web enabled. But both are labor intensive. Limited hardware oriented localized automation has been achieved at a cost without a global vision or a complete new infrastructure. In addition, costly field maintenance of hardware equipment and cash collections by attendants have remained unchanged. Revenues increase from ill-conceived, biased, and arbitrary rate rises, stricter enforcement policies, or building more parking spaces. The long-term impact is customer dissatisfaction, and unnecessary growth of parking spaces eroding urban housing development. What is needed is a national approach such as RWPTTP for uniformity. Uniformity can be obtained with a national standard of web, Internet, and online clicks. Moving away from hardware solutions with smart software can alleviate hardware incompatibilities with a more function rich platform and instant service implementations, leading to other transportation services. Let us describe some examples of progress to point to the bigger legacy system to come.

- On-street meter function emulation
- Wireless device functions emulation

- National database of parking spaces – private and public spaces

- Parking space availability database

- National vehicle database

- National rate database

- National parking policy database for uniformity

These concepts are all new in the parking industry. Proper and accurate definitions and usage are also evolving. In this chapter some logical guidelines are provided for a preliminary design. Further refinements will be required as parking standards mature.

METER EMULATION

Industry trends of late 1990s were adopted by Eximsoft to initiate this meter emulation concept. We have proven that all functions of on-street hardware meters, irrespective of their sophisticated functions, can be embedded in a software. The example is the cashless and meterless parking reservation system developed by Eximsoft based on a patent application (#US20040068433 A1) in 2002. This was the first attempt at one unified and integrated application for meter space, non-meter space, garage, lot space, and airport parking with the latest technologies then available. The same application had elements to be activated for truck parking stops along the interstate corridors in coordination with weighing stations to manage commuter traffic congestion during office hours. Elements of the patent described below cover different parking scenarios. For the application to work, each space is distinctively identified with block number, street name, deck level, and space number. It was a parking application without a smartphone for enabling centralized but integrated parking reservation for a variety of spaces, payment, and enforcement, comprising of a central computer communicating with user terminals and service terminals over a data network. There were several basic objectives behind the application development:

- Introducing a registration system for an account so that prepayment of service and account update are possible by the parker

- Testing customer convenience with a new experience of reserved or guaranteed parking space

- Supporting the 'Go Green' movement to reduce city traffic congestion and hence greenhouse gas (GHG) emissions

- Increasing the operator's flexibility for enforcement with hand-held device, statistical data collection, and report generation

- Reducing the owner's capital cost and field maintenance overheads

- Introducing an open system deviating from traditional proprietary solutions – reduction of procurement time and cost for reconfiguration as and when needed

- Verifying a more software oriented system for quick customization and moving away from a hardware-weighted system

The application pointed to a data oriented solution for the parking industry with the help of digital geographical location. It used digital location, vehicle license plate data, vehicles', and owners' information data. When a vehicle is parked all these data are checked for the authentication of auto verification and auto invoicing. Such correlation is particularly useful for securing parking areas. That is how Eximsoft obtained a trademark, a Trusted Parker Program – TPP™. A subscriber accessed the system using a wireless or wireline web-enabled terminal, or with the help of an agent, to identify a convenient, unreserved parking space, to reserve that space (advance or on the spot reservation) and pay a fee corresponding to the time the space was used. For the subscriber's convenience, payment was cashless, and was tailored to the subscriber's needs. Overtime parking protection was also available with real-time automated step-up rates instead of tickets, optimizing enforcement resources. The service terminals enabled automation in most parking operations. Efficient supervision and enforcement operations of parking facilities were ensured through the use of fixed or portable license plate readers (LPR). The parking supervisor/enforcer was provided with the ability to upload into the system the instant parking occupancy, download maps with the expected occupancy, and issue parking tickets to offenders. The information on delinquent cars could be sent to the local authority or Department of Motor Vehicles (DMV) to identify repeat offenders for appropriate enforcement in traffic courts. Since its development, there has been tremendous progress in technology for leaders and decision makers to think about consolidating parking and the transportation services. All these paradigm changes will be discussed in following sections. The objective is prepare the background materials for the next step of the intelligent and integrated transportation system (IITS); I call it the Robust Web Parking, Trucking and Transportation Portal (RWPTTP).

BACKGROUND BEHIND SMA™

The concept of SMA™ started to crystalize slowly with the development of an online mobile e-commerce transaction application. The first objective of the mobile e-commerce application was to enable Ericsson's mobile phones to make secure Internet payments while purchasing online products. The SMA™ development story by Eximsoft International is described here to help inspire new entrepreneurs in the parking industry. It was the product of intensive research by three independent companies. You can call it a strange coincidence of three companies starting with letter, 'E'. They are Eximsoft, Ericsson and Entrust. I call it power of E³. In 1999, Eximsoft was a startup company focusing its core business in system integration. Ericsson was an established company in wireless infrastructure and mobile set manufacturing. Entrust was an Internet security company developing Public Key Infrastructure (PKI) enabled embedded computer devices. All three companies were looking for killer applications using mobile sets. Nils Rydbeck, the father of Ericsson's mobile set business was convinced then that games and ring tones were not a long-term sustainable business concept.[1] In the late 1990s or early 2000s, there was an explosion of the cellphone market at the expense of landline market. Along with it came many functions and features including the electronic game capability in the cellphone. The driving force behind such dynamics was the mobility of electronic games market from table top devices. Nokia, Motorola, and many other companies were competing to capture the market share. Some experts in the industry debated the future of mobile phone set market functions development. Ericsson Research & Development (R&D) group in Research Triangle Park (RTP), NC, was supporting a protocol called Mobile Electronic Transaction (MeT) to enable mobile sets for financial transactions or mobile e-commerce (shopping cart and payment application).[2] Eximsoft secured a contract from Ericsson to develop a prototype of MeT. Eximsoft realized that

---

1 'Nils Rydbeck on Board: Former CTO Ericsson Mobile Phone Joins Opera Board of Directors', press release, June 17, 2004.
2 Global Mobile Commerce Forum, Ahonen, Tomi (February 5, 2009). 'Update on the mobile phone check-in: Finnair finds half of passengers using it'. Communities Dominate Brands. Retrieved May 6, 2017.

security of mobile e-commerce was a concern in the industry then as it is now. Ericsson was prudent enough to enable one of their mobile sets with embedded Public Key Infrastructure (PKI) capabilities of Entrust (www. entrust.com) for person to person (P2P) or business to person (B2P) or person to business (P2B) or even business to business (B2B) secured transactions. Merchant, payer, and payee would set up secured keys before exchanging electronic transactions for the specific application of their choice.

A User View Description of SMA™

Thorough research was conducted of the existing parking industry. Many challenges were observed and attempts were made to address them with innovative software solution. For example, challenges had been grouped in different categories. Categories were:

Extensive Hardware Oriented – The systems installed were mostly vendor-specific with proprietary implementation. Interconnectivity of different vendors' products was difficult, because none of them had networking interfaces and none of them was web enabled.

Capital Intensive – To start the business, the initial investment was very large. Debt services had been expensive as well. Procurements and reconfigurations were time consuming. Inventories for spare parts were expensive.

Expensive to Operate – The infrastructure needed spare parts for field services. A malfunctioning of the equipment and non-production of parking tickets even for a short period could cause revenue loss.

Security Concerns – There was no way to have details of the customer who was parking in the lot or the meter space. After the Oklahoma bombing and 9/11 terror attacks, securing the parking areas became a major concern for sensitive government locations such as airports, railway stations, seaports, and federal government's offices.

Punitive – Customer service especially on street meters was always punitive. Regulations were biased towards the owner or operator. Customers had bad experiences of being criminalized.

Lack of Uniformity and Standard- No national industry standard or uniformity had been seen in the industry. Parkers had to be oriented for every parking location. Moreover, interoperation of products was almost impossible, requiring lots of time at high cost. Parking to car ratios are close to three to one but finding a spot is very difficult due to the lack of online information.

Not Friendly to Technical-Savvy Generations – Entry and exit were not fast enough (due to Stop & Go) to satisfy the taste of tech-savvy Generation X or Generation Y. Online information about rates and availability was not easily achievable. Eximsoft wanted to start a parking industry revolution with the then available technologies such as Internet, wireless, registration to set up an account, and Interactive Voice Response (IVR – XML). Paradigms defined and developed were mostly software oriented. The first generation of software covered some aspects (given below) but more would be required as the technology advanced, the participant gained experience, and the market dynamic changed for integrated transportation services.

Use of Internet – One integrated application was developed for all parking scenarios: on-street or off-street, or open lot for a special event. One could access the web via the Internet browser to check different activities by all participants having authorized access to the system. The Interactive Voice Response (IVR-XML) system (by calling an 800 number) was used to register, activate, and deactivate parking using cellphone, replacing expensive street meters (coin or pay & display machine) or Stop & Go barrier for the off-street parking or for new spaces.

Auto Identification of Vehicles – All vehicles are pre-registered for an account using a registration form. The cash handling on the spot was eliminated by this process.

Guaranteed Parking Space – Circling for an open space was avoided by reserving the space prior to going to the place (be it on-street or off-street). This helped CO_2 reduction. Traffic congestion was also minimized.

Decriminalizing Parking Services – Civility was disrupted by confrontations with an enforcement officer (EO) writing a penalty for an over stay. Instead, real-time step-up rates were introduced where the parker is charged higher rates. By this process, congestion in the traffic court was reduced with increased revenues in many cases. It also minimized efforts to chase uncollected fine revenues.

Floating and Home Printed Permits – Spot-specific on-street parking was avoided by a floating permit. You print out a daily or hourly city permit to park anywhere as long as there is a free space. In addition, an off-street printed permit after you made a prior reservation allowed you to park in a spot of your choice.

Online and Real-Time Enforcement – Enforcers could download the latest information of occupancy in their hand-held device for customers who had not paid. They could write the ticket or accept payment on the spot as appropriate. Enforcers did not have to walk to every car because they had the prior information about occupied spaces by block, street, or parking space number.

Faster Dispute Resolution – All data of ticketing are recorded online and real time with time stamps and photographs. Parkers could also file complaints online and real time. Disputes could be resolved online to all parties' satisfaction with exchange of electronically documented proof.

Account Management by Parkers – Parkers' activities are recorded, including their payment history. They would be notified of their account status. They can print out their parking payment receipts for reimbursement if required.

The rationale behind such progress was a new service supported by software changes. A strategy was developed to list key elements shown below that could have been impacted by software. Each of them was thoroughly analyzed so that software modules developed were capable to handle them in isolation or in a combination as needed.

No Meter – Saving millions of dollars in capital investments depending on the configuration.

No Pay & Display – No need to walk from the Pay & Display machine during rain and snow to get a permit.

No Vandalism – Meters are subject to vandalism in the field. You lose revenues once the meter is non-functional. Software does not have any element in the field to be interrupted for service.

No Spare Parts – No field maintenance or reconfiguration, saving time and resources.

Scope for Creating New Service – Open platform for software dependent services by small code modification in a short time rather than vendor dependent hardware.

No Cash Collections – Online or pre-payment helped to avoid cash handling and associated labor costs – a major cost element of the current platform.

Scope for Defining a Class of Service – Software modules helped to make a service differentiation to attract different kinds of customers for convenience. The software platform also creates an option for on-fly service implementation.

As explained above, the Cashless Meterless Parking & Reservation (CMP&R) System of Eximsoft was an overarching solution for the Parking Industry. The integrated system was applicable for all the following parking scenarios:

- **On-street parking**
- **Off-street parking**
- **Event parking (e.g. sports arena/concert venues) – with premium reservation options**
- **Airport parking – with premium reservation options**
- **Truck parking** – The software application developed was so robust and powerful to cover the truck parking along interstate highway corridors, reducing traffic congestion during office commuting hours. The process helped coordinate activities with weighing stations for resource planning and drivers' schedules.

Each of the above parking scenarios requires only a simple, one-time prior registration with the service provider by the end user. The objective of registration is to set up an account with an access code for identification from anywhere. Users can access the system and its services from a variety of locations and terminals:

- **Wireline phone**
- **Cellular phone**
- **Web-enabled wireless device**
- **Web terminal**
- **PC**

The registration form is depicted in Figure 4.3. The registration captures user details like user name, mobile phone number, preferred mode of payment (credit, debit), and the vehicle detail. Using this facility, the user can create or modify his/her existing profile. Once the registration

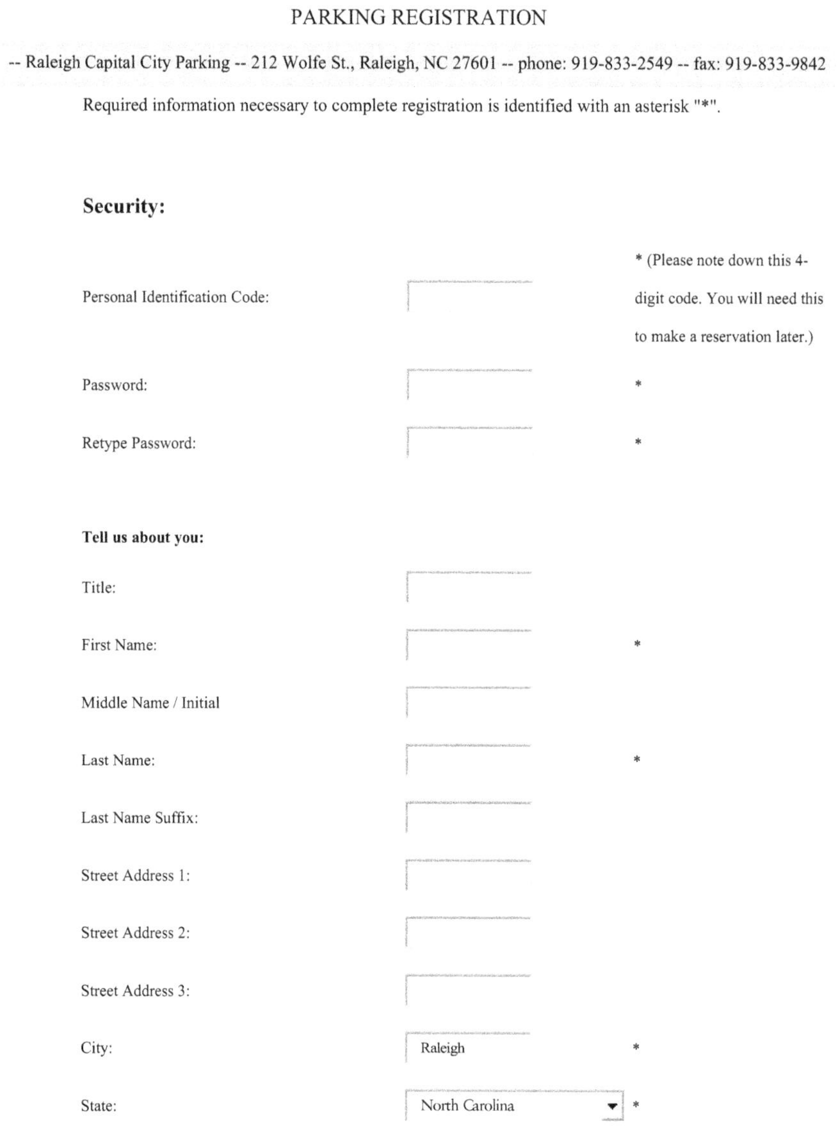

PARKING REGISTRATION

-- Raleigh Capital City Parking -- 212 Wolfe St., Raleigh, NC 27601 -- phone: 919-833-2549 -- fax: 919-833-9842

Required information necessary to complete registration is identified with an asterisk "*".

Security:

Personal Identification Code: _____   * (Please note down this 4-digit code. You will need this to make a reservation later.)

Password: _____   *

Retype Password: _____   *

Tell us about you:

Title: _____

First Name: _____   *

Middle Name / Initial _____

Last Name: _____   *

Last Name Suffix: _____

Street Address 1: _____

Street Address 2: _____

Street Address 3: _____

City: Raleigh   *

State: North Carolina ▼   *

FIGURE 4.3   (Continued).

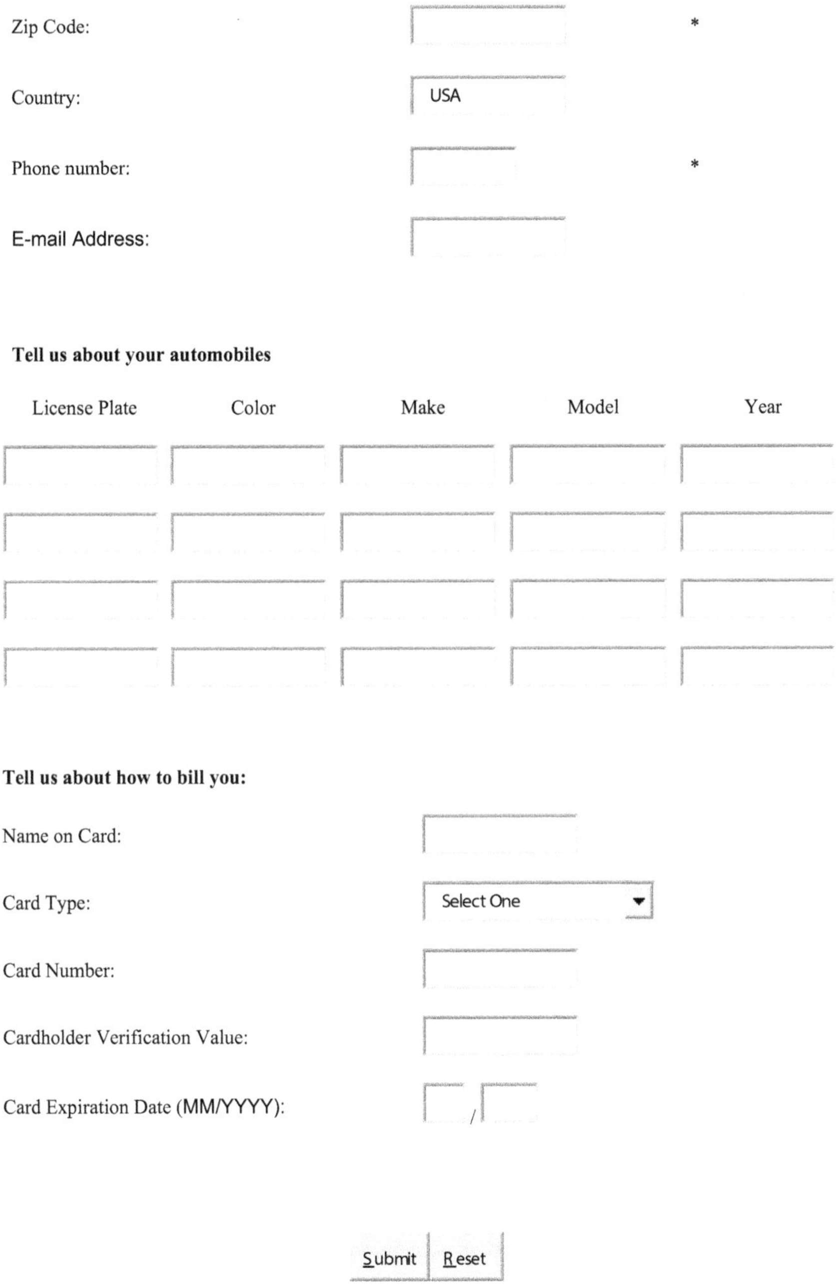

Zip Code: [   ] *

Country: [USA   ]

Phone number: [   ] *

E-mail Address: [   ]

Tell us about your automobiles

License Plate	Color	Make	Model	Year
[ ]	[ ]	[ ]	[ ]	[ ]
[ ]	[ ]	[ ]	[ ]	[ ]
[ ]	[ ]	[ ]	[ ]	[ ]
[ ]	[ ]	[ ]	[ ]	[ ]

Tell us about how to bill you:

Name on Card: [   ]

Card Type: [Select One   ▼]

Card Number: [   ]

Cardholder Verification Value: [ ]

Card Expiration Date (MM/YYYY): [/]

[Submit] [Reset]

FIGURE 4.3   Registration form.

process is successful, users are given a PIN code. Further customization of this form was also possible if requested by specific client.

The CMP&R system exploited distributed network elements or devices over the Internet for connectivity. Mobile devices were also connected for a robust innovative network to complete all functions of parking. For example, capabilities achieved from such a network are:

- **Secure web connectivity**

- **Scalable robust web servers**

- **Web Browser access – PC or cellphone**

- **Interactive voice response – and, eventually, voice recognition server**

- **Different databases**

- **Authentication services**

The network architecture shown in Figure 4.4 was used for initial end user registration, parking fees collection, premium reservation, activation and deactivation of parking, priority access to a parking lot, as well as parking enforcement and enhanced security. At the core of the CMP&R system processes reside the novel and innovative ways of integrating a variety of technologies and Internet services to achieve our objectives:

- Computer server technology with intelligent agents

- Geographic information systems (GIS)

- Secure internet web services

- Geographic position system (GPS)

- Wireline and wireless communications networks

- Desktop computers

- Web enabled mobile communication devices

- Secure internet payment systems

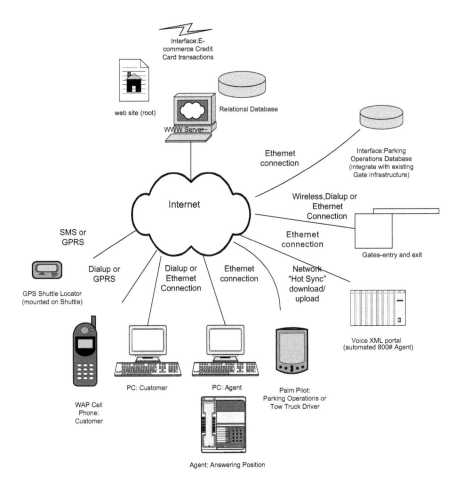

FIGURE 4.4   Network architecture.

- Parking paradigm folded under information technology (IT) domains

- Digital signatures and identification systems, PKI and SIM cards.

Figure 4.5 shows all functional modules developed to integrate meter and garage parking applications with minimum hardware interfaces – no meter, no barrier, and no ticket-spitters. The configuration and the space management module were developed to configure the space without the hardware but with data such as space number, street address, or other needed parameters from the operator. The platform module constituted

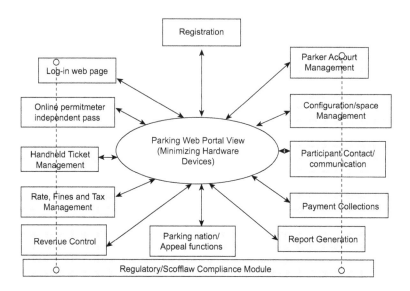

FIGURE 4.5   Software modules for CMP&R or SMA™.

the basis to connect other functional modules with needed application program interfaces (APIs). The platform module is what glues in all other modules, including complying with regulations and dealing with scofflaws. The user interface started with the registration input from the customer of the web portal. After the registration, the user could log into the account created with an appropriate user identification and password. Functions and features achieved through these different software modules are:

- Registration and its confirmation process
- Reservation confirmation process
- Bar coded permit printing
- Account management process
- Online enforcement and data collection for dispute resolution
- Account consolidation
- Interactive voice response
- 800 call for street parking and garage parking

- Interactive voice response to activate or deactivate

- Mobile identification generation

Additional modules had been developed and tested for the extension of the application for the truck parking solution in the lab. It included many functions specific to truck drivers' sleeping and driving schedules, truck routes, and associated state by state regulations so that a compliance remotely and online was possible.

STREET PARKING CASHLESS METERLESS PREMIUM AND LEGACY-COMPATIBLE SERVICES

The user in his/her car finds a suitable (permitted, marked, and uniquely identified as such) street parking spot. The legacy service would require him/her to insert coins in the meter (if available) to secure a finite-time parking allowance, or go to the nearest ticket-dispensing machine to acquire a time-predetermined parking voucher to be displayed on the car dashboard. While such legacy services may continue to be available, in the presence of the Eximsoft CMP&R system, the user has the option of securing a premium cashless parking service by using a web-enabled mobile device to access the CMP&R system or simply by dialing an 800 number using the cellphone or a wireline phone to connect to the parking system. Through its interactive voice response features, the CMP&R system prompts the user for PIN code (in certain circumstances the PIN code may not be necessary as the cellphone number itself can be the unique identification of the user) and the parking spot unique identifier (e.g. meter number or physical parking space number). In the absence of a unique parking space identifier the user provides the system with just the license plate number. The CMP&R system immediately validates the input given by the user and activates the parking service. It also gives the user confirmation of parking at the specific space. To deactivate the parking the user calls the same 800 numbers and just enters the PIN code and the space number or meter number, following which the CMP&R system deactivates the parking with a message to the user of time spent and the amount charged to his/her credit card or prepaid account. The system also sends the parking activation and deactivation notifications online to the parking enforcement authorities via their wireless hand-held devices in order to inform them and to facilitate their enforcement functions.

Challenge to the Application

It was progress, no doubt, but there are more challenges to overcome for the real implementation of RWPTTP. Parking, especially street parking, over the years had become a key factor in the economic growth and image of cities. As a result, it has become a political issue for the mayor and city council members to collect more revenues and to divert funds from parking revenues to other city budgets. Challenges to the application were to have a transition time to replace thousands of old meters, contain new capital investments, lowering the inventories, less ticketing to displease city dwellers and visitors, and to minimize field maintenance including cash collections. CMP&R or SMA™ had addressed all the issues one by one to bring a paradigm change by designing a web-based solution with capabilities of e-payments, online problem resolution, e-citations with step up rates instead of tickets. In addition, it was possible to decriminalize the parking industry, enhance the business process online, link to Department of Motor Vehicles (DMV), if needed, to report stolen vehicles or other criminal activities. Technologies are reaching at different maturity levels, requiring ongoing strategies of improvement.

GARAGE AND CAMPUS PARKING

The user enters the garage by using the existing system and its legacy infrastructure. The process for activation and deactivation of the parking is the same as the street parking, with the user providing the PIN code number and the license plate number to the system. The system will send the user license plate number online to the wireless hand-held device and/or to the desktop terminal of the garage attendant at the exit gate. When exiting, the user gives the attendant the ticket and the license plate number. The attendant registers the time and verifies the match of the license plate number with the system record. The gate will open on the reconciliation of the license number with the parking record, and the user will be debited with the corresponding parking fee. An automatic license plate reader or user card identifier can be incorporated to further automate the process. For new garages, the process was much simpler – just use the reserved lot number and the space number received during pre-registration. Event and campus parking had the same procedures.

Challenges to the Application

The parking challenges for garage and campus parking had been studied for years with no historical data collection in place. A rule of thumb approach, due to the lack of historical data, led to parking spaces taking over the commercial or recreational real estate of a city. If this continued there would be hardly any space left for the economic development of cities. Challenges were regulations on number of parking spaces per square feet of retail and office spaces, hourly rate, quantities of residential permits, ticket and permit processing, event parking, and the utilization of spaces for an optimum revenue outputs. Manual operation could have been replaced with semi-autonomous procedures in most cases of a large campus with 10,000 annual permits, 55,000 daily permits, 350 meter spaces, and 35,000 annual parking tickets. The opportunity of defining different classes of services to earn more revenue opened up based on proximity criteria and pre-payment options. The same web solution was used as in the street parking with some added functions and features specific to garage or lot parking with no barriers. Cash attendants were replaced with easy to enforce parking violations and on the spot revenue collection, if needed. But much more online capabilities are required to develop user-friendly functions for the decades to come.

EVENT PARKING

The event parking process starts with the one-time user registration with the event parking authority, largely as described in the street parking scenario. The user accesses the system via a web-enabled device (PC, PDA, or webphone), or he/she dials the 800 number using a cellphone or a wireline phone to connect to the parking system via voice to machine language (XML). The CMP&R voice response module prompts the user for the event (date and/or time) together with the user PIN code (in certain circumstances the PIN code may not be necessary as the cellphone number itself can be the unique identification of the user). The system immediately verifies the input given by the user and initiates a parking reservation. Optionally, the CMP&R system may offer a premium service related to priority parking location (a feature much appreciated in inclement weather and/or at large sport events). The user is given a message by the system confirming the parking reservation for the event with a unique parking authorization code (optionally identifying the premium parking spot reserved), together with the amount charged to the user credit card.

The unique authorization code given to the end user is also downloaded to all the enforcers' hand-held wireless devices on the day of the event. At actual parking time, the parking attendant/enforcer will match the authorization code against the record in the hand-held device and allow the user to enter the parking lot. Optionally, the attendant may direct the user to the premium reserved spot. Further virtualization of parking and transportation services (autonomous vehicles) may eliminate these options eventually.

AIRPORT TERMINAL PARKING

Large airport parking is a very special case different from other garage parking because in big airports there are shuttle services to bring people back to the terminal. Here, people's convenience and time are critical so that they do not miss the flight. Provisions for hourly, daily, and many days were required in addition cellphone parking to pick up passengers. The fundamental process for the end user starts with the one-time registration with the Airport Parking Authority as in the case of street parking above. The Airport Parking Authority will provide a well-marked reserved area in the existing parking garage or the lot, which is at close proximity to the departure gates. This is a premium parking service. The user accesses the Airport Authority CMP&R system via a web-enabled device or by dialing the 800 number using the cellphone or a wireline phone. The CMP&R system handles the web access or uses its voice response module to prompt the user for his/her PIN code (in certain circumstances the PIN code may not be necessary as the cellphone number itself can be the unique identifier of the user) and the dates when the reserved parking is required. Optionally, the user may be prompted to provide the airline flight, such that the system can offer the most advantageous parking spots as a premium service. As mentioned above, the user can avail of any web-enabled device (PC, PDA, webphone) with Internet connection and do the parking reservation via the Web. The system immediately validates the input given by the user; the user receives a message from the system confirming the parking reservation for the dates and the amount charged to the user's credit card for the specifically reserved spot. The user enters and exits the airport parking facility using the existing system and pays the regular parking fees. The license plate number of the reserved parking user is downloaded to the wireless hand-held devices of the parking attendants as required to facilitate enforcement functions. Further graceful integration of a light rail system may add value to future automation.

Challenge to the Application

Airport parking services had to compete with many adjacent private parking facilities. That is why service differentiation was a must. It was done with innovative programs such as pre-paid guaranteed parking space, home printed parking pass with bar code, easy access without barrier, loyalty program for repeat customers, enhanced security for parkers, cellular customers waiting area to pick up, and uniform and standard national parking paradigm for convenience of business travelers. The project success was measured in the economic scale – scope of new revenues and cost savings in both the capital investment and yearly maintenance. Further improvements may be on the horizon.

Get a preview of cost elements for those who feel uncomfortable with emerging technology. A piecemeal solution with transitional approach may not provide the best economic incentive but a grand scheme may work better as will be discussed in a later chapter. A typical scenario of a city with 3,000 meters was analyzed. As shown in Figure 4.6, replacing meters gave opportunities to save hard costs in many areas such as unit cost, installation of meters, system integration, systems maintenance, and cash collection from each meter. For meter collections, there were many interesting stories – there might not be enough money in the meter to pay $15 per hour labor cost. In that situation, your operating costs would go up linearly as you kept adding meters. In a meterless scenario, you can add more spaces without any extra

	Hard Meter	Soft Meter	Saving By Unit	Total Savings for 3000 Meters
Unit Cost	$500	$50	$450	$1.5 million
Installation	$150	$0.00	$150	$450,000
Systems Integration	$100	$20	$80	$240,000
Systems Maintenance	$75	$10	$65	$195,000
Collection Cost Per Meter	$50	$0.00	$50	$150,000

FIGURE 4.6   Saving opportunities with SMA™ for street meter.

costs – a question of minor change to software codes. Such a reverse relationship helped profitability. According to national records, there are close to 5 to 6 million meters in US streets, resulting billions of dollars in cost savings. Savings per year could cover new technology costs with a pay-off period of less than a few years. Almost all existing parking hardware functions had been emulated with software innovations. The weighted average cost of these software developments was about $50 per meter space for the 3,000 meters base. The weighted development cost could have been lower with a larger base (more than 3,000 meter spaces). There were other two differential cost elements for such a parking environment: system integration and yearly operation. Both were much lower than the hardware configuration because they became a part of the information technology (IT) infrastructure. Other tangible benefits were no vandalism of hardware resulting in revenue loss and inconvenience of customers to walk to/from the sophisticated Pay & Display machine during the adverse weather condition.

The cost/benefit analysis for airports was carried out in terms of reserved parking as shown in Figures 4.7 and 4.8. Figure 4.7 indicates the daily airport rate for different parking configurations and the availability of total spaces in fourth quarter of 2003. Figure 4.7 indicates the possible increased revenues for reserved spaces (called premium spaces) at $5 premium rates over normal rates. A conservative number of 5% of total spaces had been assumed to be premium spaces assigned to business travelers only. The last two columns of Figure 4.8 show the increased revenues to 90% and 80% occupancy rates. In either case, the pay-off period of the technology costs is very short. In fact, because of high demands by customers, many airports had offered such services and its popularity went up since 2003 even at a higher than $5 per hour rate in addition to normal rates depending on the distance of the reserved parking lot from the airport terminal.

In addition, there was another subjective measure called customers' convenience for a possible future paradigm revolution. Opinions of 80% to 90% of few hundred customers surveyed locally were in favour of these changes. The global benefits for Integrated Parking Scenarios were:

- One operation for all parking scenarios
- Capital containment for both on-street and off-street parking configuration and maintenance

City	Airport	Daily parking rates					#parking spots (000's)
		Valet	At terminal	Shuttle to terminal			
Atlanta	ATL	--	$12	$9	$8	--	30
Chicago	ORD	$32	$25	$13	--	--	21
Los Angeles	LAX	--	$30	$10	$8	--	8
Dallas / Fort Worth	DFW	--	$13-$16	$9-$10	$6	--	28
Phoenix	PHX	--	$16	$5	--	--	8
Denver	DEN	$24	$7	$5	--	--	4
Houston (George Bush Int'l)	IAH	--	$12	$11	$8	$6	23
Las Vegas	LAS	$14	$8-$12	$8	--	--	15
Minneapolis / St. Paul	MSP	$20-$25	$12-$14	--	--	--	18
Detroit	DTW	$26	$10	$8	--	--	20
San Francisco	SFO	$38-$45	$22-$35	$11-$13	--	--	14
Newark	EWR	$28	$20	$10	--	--	14
New York City (JFK)	JFK	--	$15	$10	--	--	14
Miami International	MIA	--	$10	--	--	--	13
Seattle-Tacoma	SEA	$30	$20	--	--	--	12
Orlando	MCO	--	$15	$7	--	--	15
St. Louis	STL	--	$18	$11	$8	$7	12
Philadelphia	PHL	--	$17	$9	--	--	11
Charlotte / Douglas	CLT	$19	$6	$3	--	--	11
New York City (LGA)	LGA	--	--	$15-$30	--	--	11

FIGURE 4.7 Top 20 US airports with parking rates and available parking spaces (2003).

City	Airport Code	Total parking spaces	5% are "premium" spaces	$5 per day per user	
				Annual revenue - 90% occupancy (millions)	Annual revenue - 80% occupancy (millions)
Atlanta	ATL	30,000	1,500	$2.46	$2.19
Chicago	ORD	21,000	1,050	$1.72	$1.53
Los Angeles	LAX	8,000	400	$0.66	$0.58
Dallas / Fort Worth	DFW	28,000	1,400	$2.30	$2.04
Phoenix	PHX	8,005	400	$0.66	$0.58
Denver	DEN	4,000	200	$0.33	$0.29
Houston (George Bush Int'l)	IAH	23,000	1,150	$1.89	$1.68
Las Vegas	LAS	14,043	702	$1.15	$1.03
Minneapolis / St. Paul	MSP	18,000	900	$1.48	$1.31
Detroit	DTW	20,000	1,000	$1.64	$1.46
Orlando	MCO	14,500	725	$1.19	$1.06
Tampa	TPA	~10,000*	500	$0.82	$0.73
Milwaukee (General Mitchell)	MKE	~5,700*	285	$0.47	$0.42

FIGURE 4.8 Expected revenues for airport reserved parking.

- CO_2 emissions and traffic congestion reduction

- Enhancing parking area security

- WiFi networking for real-time online information

- Creating smart city environment

- Reversing the trend of cost increases for changed and new configuration

- Becoming part of information technology (IT)

- Introduction of different classes of services catering to customer needs

- Decriminalization of services and traffic management

- Instant rate change and enhanced revenue management

- Replacing arbitrary rate increases with a well-defined class of services for convenience

TRUCK PARKING ON INTERSTATE HIGHWAY CORRIDORS

Transportation of goods by trucks cannot be overlooked. Its importance has gone up further due to the fact that smaller autonomous trucks can take over the transportation of goods door to door (last mile) in the next decade and reduce the competition with daily commuters in traffic jams. Three areas of the truck industry were further investigated to examine the suitability of the software program for the truck parking industry as part of RWPTTP implementation. The key finding of the investigation was that there were not enough parking facilities along the interstate highway corridors. Developments of such parking facilities could add rural and urban cooperation, economic development in addition of improving the following for the truck industry:

Truck Operation Improvement – Technologies such as automatic vehicle identification (AVI), bar-coded permit, and WiFi connectivity of mobile devices would improve the daily truck operation across interstate highways as well as across the borders of Canada and Mexico.

Truck Traffic Management Efficiency – Online information of truck size, load, types, etc. could be broadcast to the interested parties as the truck moves from state to state on highways. Driver's schedules and routes, including truck maintenance records according to each state's regulation, could be checked online prior to reaching the check points.

Road Safety Enhancement – Truck parking close to the weighing stations could be coordinated well to improve safety, checking hazardous materials and weigh station resource coordination. In addition, commuter truck entry to the city could be managed well to avoid office commuter traffic if appropriate parking facilities are located close to the weighing stations.

Materials presented above paved the way to use the cellphone and its mobility to make a first step forward for enhancing the existing parking paradigm. The smartphone and its robust application capabilities with the latest technologies will be discussed in accompanying sections. The future infrastructure would eliminate the need for sophisticated but expensive street meters and barriers in the parking industry in addition to causing a total overhaul of transportation services. Many technologies were left out while developing the above application because they were not fully mature at that time, such as drones, autonomous vehicles, geo positioning systems (GPSs) Bluetooth light emitting (BLE) beacons – just to name a few. In addition, the integration of transportation and parking services were not given the due consideration they deserve. Such consideration would have revolutionized all of autonomous transportation services. The National Parking Association (NPA), International Parking and Mobility Institute (IPMI), North American Transportation Services Association (NATSA), American Trucking Association (ATA), Conference of Minority Transportation Officials (COMTO), and other international bodies have since explored these technologies. Let us review some of those for new economic development. They are highlighted to generate interest among decision and policy makers on how to overhaul parking and transportation for the next century.

MOBILE DEVICE FUNCTIONS EMULATION AND PARTIAL AUTOMATION

Since the development of SMA™, application development efforts have increased and also technology has advanced. Developments of mobile payments are noteworthy for both. The good news is that such efforts, in mixed configurations, can emulate the present parking hardware – meter and gate. Such devices can even replace them during the transition period from manual to semi-autonomous. One such example was developed and patented by Transparent Wireless Systems, LLC (www.transparentwireless.com/). Much of this was taken from a report (dated April 17, 2017) written by Santanu Dutta, Ph.D., founder of TWS. Dr. Dutta focused on improvement of mobile payment from the installed mobile payment – a preamble of future automation of integrated parking and transportation services. Examples of those technologies are BLE beacon, NFC, smartphone, precision GPS, different satellites, drones. Hopefully, such robust functionality, connectivity, and built-in intelligence can be leveraged to significantly enhance user experience in both on-street and off-street parking systems.

Preview of Mobile Payment for Parking

Pay-by-cell has begun to address on-street parking only because of quick payoff, but has barely scratched the surface of off-street (gated garage) parking which is a challenge due to many technological obstacles. Marketing efforts by Parkmobile, PaybyPhone, and others in addressing on-street parking has demonstrated a value of pay-by-cell to all stakeholders.[3] This value resides primarily in the *user experience*, i.e. in enhanced user convenience with no change in the infrastructure. This windfall can be leveraged to significantly enhance user experience in both on-street and off-street parking. This chapter describes some examples of how emerging technologies may be exploited to substantially retrofit both on-street and off-street parking payment systems only.

Deployed Wireless Payment Systems

Over the past several years, a number of companies have had limited success in offering cellphone-based payment systems (commonly referred to as 'pay-by-cell') for street parking. It is an interim solution because it does not explore the potential of on-the-horizon technologies. Present pay-by-cell offerings for street parking follow a fairly standard pattern, consisting of the steps described below. Some service providers offer two versions – one based on an app downloaded to a smartphone and the other based on answering automated, telephonic voice prompts. The latter is not very convenient if it has to be performed on a busy and noisy street. Reports suggest that the app-based service is overwhelmingly

---

3 In Washington, DC, more than 50% of on-street parking revenue is collected by Parkmobile's pay-by-cell system.

more popular than voice-based pay-by-cell. Typical steps in present pay-by-cell parking payment are described below.

Registration
The user registers for service with the service provider. Besides ID information required for any internet payment transaction, such as name, address, and credit-card/bank-account details, the subscriber also registers the IDs of the cars (license plate numbers) and mobile phone numbers, which will be used for user authentication by the Payment Application Server.

Start Parking
A parking session is started when the user has communicated the following information to a server on the web (Payment Application Server) operated by the Payment Service Provider (all of the companies mentioned above are such providers):

a) Parking location (this determines the tariff, or parking rate). In most systems today, this involves the user reading a posted zone ID number (often pasted to a legacy coin meter) and entering it on his phone. Relatively recently, QR Quick Response) codes have been posted instead of a zone ID, which the user scans with his smartphone camera. Clearly, having to enter a zone ID number was recognized as a negative factor for user experience.

b) User ID. The phone number serves as the user ID in all systems. A password is required for registration and for accessing the service provider's website but, in most systems, users are allowed to store the password in the phone to avoid entering it each time that a parking payment transaction is executed. In this way, launching the app from a registered phone automatically submits the password for authentication by the server, transparent to the user, and takes the user straight to the parking payment application. This is not the most secure method of transaction available today but, given the typical payment amounts involved, is generally accepted as sufficient for risk management.

c) Vehicle ID. This is the car's license plate number. In the worst case, this has to be entered manually by the user but all systems allow the last used ID to be used as the first default.

d) Duration of stay. The need to communicate this is determined by parking authority policy. In some jurisdictions, such as Washington, DC, the user is required to commit to a predetermined parking session time at the start of the session, as is required for coin meters; in others, the user does not need to select a predetermined session time and pays only for the actual duration of stay. The need to select a parking session time adds a step of user interaction in all systems, which is a negative from a user experience perspective. Some US cities do not require advance payment for an estimated parking session time. In such cases, this extra step can be eliminated.

Once the above information has been received by the Parking Application Server, it authenticates the user and sends a confirmation of the start of session. All systems allow the user to view the status of an active (open) session remotely on her/his phone, and to extend it if necessary. Remote session time extension has been one of the major end user benefits of pay-by-cell parking payment. Another feature, also supported by most systems and considered important by city parking authorities, is displaying the present parking rate and terms of occupancy, such as maximum session time, free time, weekday/weekend differences, and others. These policy statements are conveyed to the user *before* the session is initiated, allowing him to abort the transaction if he finds the terms unacceptable. This has gained importance with the recent introduction of dynamic, demand-based parking rates by some city parking authorities. In fact, this is viewed as one of the major benefits of pay-by-cell parking payment from the perspective of city authorities.

End Parking

Not all service providers allow the user to end an active parking session in all parking scenarios. In cities where the user is required to commit to a predetermined parking time, Parkmobile does not allow the user to close an active session. The thinking behind this may have been that, as the session is prepaid with no refund for unused time, it is moot whether the user closes the session before driving off. We will see that there are important enforcement advantages in incentivizing the user to close the session even in prepaid parking, or automatically detecting the end of the session if the user forgets.

Figure 4.9 shows the essential elements of an improved pay-by-cell system by TWS over the existing deployed ones embodying emerging technologies all based on a smartphone app supported by both iPhone and Android operating systems. They are described here for a transitional system before the full-blown RWPTTP detail is worked out. For example, autonomous vehicles with embedded GPS functions may replace smartphones – an ideal future configuration for no meter and no gate.

Improvement Areas

The user creates an account with the parking Payment Service Provider via a registration process, during which he/she provides the usual credentials required for any payment app. These include a credit card number or bank account and the IDs (license plate numbers) of the vehicles to be used. The user account is best created by the user prior to parking at his convenience rather than at the time of parking, although some systems allow both registration and payment transaction via voice prompts. The wireless device is almost invariably a smartphone, as the

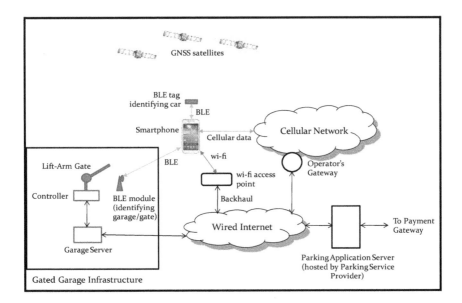

FIGURE 4.9 Essential elements of a smartphone app based parking payment system.

economics of most parking payment business models do not support a dedicated wireless device such as RFID tag today with sufficient functionality and battery drain characteristics. The present model for highway toll collection has a dedicated RFID tag (either passive or semi-passive) issued to the user by the toll collection agencies. In some cases, automation is achieved without the RFID tag by high-speed cameras over the highway to capture the license plate as vehicles pass under the camera. The license plate is then matched with DMV databases (state or out of state) for addresses and an invoice is sent via regular USPS. The interest in dedicated, smart parking tags in the car may change in the future, leveraging IOT technologies being developed for 5G cellular systems,[4] but is not there today. Autonomous vehicles or vehicles of next decades will be IOT compatible, automating some of steps described during the transition requiring registration for identification and the invoice preparation. Drones are already IOT compatible. Using drones shown in Figure 4.10 for

FIGURE 4.10 IOT compatible drone for traffic management (stationary or non-stationary).

---

4 One class of IOT devices (referred to as Massive IOT) are expected to have battery life measured in years and connect to 5G cellular networks with very low throughputs.

parking enforcement and traffic and toll services will go a long way to initiate automated integrated transportation services and 5G cellular systems may serve as a web backbone of RWPTTP enhancing high-speed connectivity with mobility.

With the smartphone or autonomous vehicle as the wireless device, the connection to the Parking Application Server on the web can be achieved by any of a range of RF connectivity options available now with cellular data, WiFi, Bluetooth, and infrared. However, there is a trend to use drones with minimum power drain similar to other short-range sensor technologies like Bluetooth and NFC. Cellular data connectivity may be assumed to be ubiquitously available for all on-street and off-street parking in most markets such as airports, railway-ports, shopping centers, etc. However, cellular coverage may be limited in underground garage locations and a stationary drone may be able to overcome it. In constructing use cases for such locations before a drone is perfected for underground off-street parking, one of the following approaches may be adopted. The connectivity require-ment should be limited to the entrance and egress points, where cellular connectivity may be available. Alternatively, one of the following methods may be used.

(i) WiFi-based internet connection is provided directly to the phone, whereby the phone can connect with the Parking Application Server. If the WiFi connection is encrypted, the users will have to be registered for service beforehand. This may be a disincentive both for the end user and the garage owner.

(ii) BLE/WiFi protocol-translating access points are commercially available. These may be installed inside the garage, whereby the phone communicates by BLE to the access point, which then converts the messages to WiFi and connects to a garage-provided WiFi hub (and thence to the internet).

ELIMINATION OF INPUTTING ZONE IDS

The examples below are taken mostly from systems developed by TWS – an improvement on all other installed systems. The experiences of the user and the enforcement officer are the primary performance indicators. By utilizing the GPS and Bluetooth Low Energy (BLE) features of a smartphone, the following improvements may be made to the user experience, quantified by the number of required user interactions

FIGURE 4.11   Parkmobile Signs in Washington, DC.

with the phone. Eliminating the zone ID obviates the need for the user to input the ID, typically pasted on a parking meter, as shown in Figure 4.11. The zone ID may be eliminated by using the navigation feature[5] of the smartphone to automatically determine the location where the vehicle is parked. It may seem strange, given that GPS was available on smartphones before pay-by-cell was introduced around 2009 that the present pay-by-cell systems, with the exception of PANGO in Europe, chose not to use it. One reason may be the poor accuracy of GPS in urban areas. This is caused by multipath and blockage. Figure 4.12 shows an iPhone screenshot of a GPS indicated position in downtown Washington, DC.

---

5 Today's high end smartphones track not only GPS but also GLONASS (Russian constellation). The more constellations that a satellite navigation receiver tracks, the better is the performance in urban canyons, where some satellites of all constellations are likely to be blocked by tall buildings. Modern receivers are able to obtain a position fix from satellites belonging to diverse constellations. At least 4 satellites need to be visible for an accurate and reliable fix.

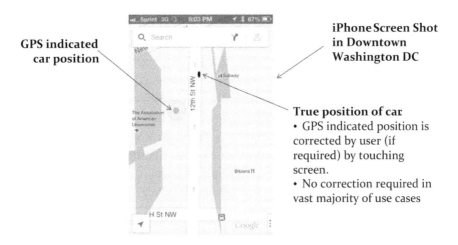

GPS indicated car position

iPhone Screen Shot in Downtown Washington DC

True position of car
• GPS indicated position is corrected by user (if required) by touching screen.
• No correction required in vast majority of use cases

FIGURE 4.12   GPS based parking location determination with provision for user correction.

SELF-CORRECTION OF USER LOCATION

Self-correction of location using BLE beacon is another example of TWS systems. The screenshot in Figure 4.12 shows an example of the inaccurate reading provided by GPS in an urban canyon. TWS describes a simple method of allowing the user to correct the indicated position by touching the *approximately* correct location on a touch-sensitive screen. The accuracy does not need to be greater than a city block as the parking rate typically does not change with finer granularity. In most cases, the user does not need to do *anything* because the indicated position is sufficiently accurate (less than 5–10 m error is common in many urban and suburban environments). However, if the screen shows the car on the wrong street or inside a building, the user can point to an approximately correct position, as shown in Figure 4.12. It will be apparent that inputting a corrected position on a touch screen is much simpler than finding and inputting a zone ID, or scanning a QR code on a legacy coin meter, as is the current practice in existing systems.

An alternative to the above, proposed in some systems,[6] is to mount a beacon (e.g. a BLE module) on the parking meter. It is not recommended because it runs the risk of interference from adjacent beacons. A better solution is a single beacon per city block, which would revert the location

---

6 Jean-Louis Jean-Louis Fiorucci et al., 'Providing City Service Using Mobile Device and a Sensor Network', US Patent 9,159,080 B2.

accuracy to that of the TWS system described above and also installation of new devices is required for the beacon solution, which is challenging in urban areas. The TWS system avoids installing any new device, beyond that considered desirable for advertising. If the TWS app were built into the vehicle's electronics, the above user correction would be unnecessary, as much greater accuracy can be achieved with a vehicle-mounted navigation system. Today, a vehicle's positioning system will typically provide approximately 1–2 m accuracy compared to 5–10 m for a cellphone with good satellite visibility. The greater accuracy of a vehicle's positioning system stems from motion sensors and is also called 'inertial assist'.

In the future, requirements for driver assistance in vehicles (leading ultimately to driverless cars) are expected to drive the vehicle's integrated GNSS system to better than decimeter accuracies. This, in combination with support for two to four satellite constellations in the GNSS receiver, will be more than adequate to identify the exact parked location of the vehicle in most urban locations – the genesis of RWPTTP's auto identification in all open parking spaces, a move to relocate garage spaces to SPDs and TPDs. The parking authority also benefits from payment systems that do not need zone IDs. Signage space is very limited on streets, which are already cluttered with many different types of signs, especially in large cities. It is clear from Figure 4.11 that there is no room for a second service provider. This may force the parking authority into an exclusivity arrangement with the chronologically first service provider, locking out the benefits of competition indefinitely. In contrast, if the parking payment system did not need to monopolize the signage space in a city, multiple systems could operate in parallel, competing for customers based on cost and quality.

SENSING OF VEHICLE ID BY SMARTPHONE

A prime example of a recent wireless technology, pulled by IOT but applicable to parking, is Bluetooth Low Energy, or BLE. Originally, it was introduced in 2010 as Bluetooth 4.0; recently an improved version has been introduced as Bluetooth 5.0. BLE has modified the older Bluetooth technology (Classic Bluetooth) in a number of ways, such as:

- Current drain was reduced substantially, making it suitable for operation with watch batteries for months to years.

- Synchronization and exchange of short messages between BLE modules was made much more rapid than traditional Bluetooth pairing and data exchange. BLE modules can sync in under 3 ms (depending on parameter settings) compared to traditional pairing, which takes many seconds.

- Profiles such as proximity were introduced, which enables one BLE device to recognize that it is close to another and engage in a rapid exchange of small messages with the proximate device.

- Reduced complexity and anticipated IOT scale lowered retail prices to a few dollars.

- Both iPhone and Android operating systems support it.

- Security was improved over traditional Bluetooth.

These characteristics make BLE suitable for many future parking applications, some of which will be useful for RWPTTP design. Such BLE beacon integration with autonomous vehicles will further simplify RWPTTP design. For example, this is of particular benefit for a rental or fleet car. In fact, in addition to payment for parking, a BLE based association of a vehicle with a smartphone could facilitate the issuance of rental cars to customers and subsequent rental payment. This could materially improve the user's car rental experience.

PARKING SESSION EXTENSION AUTOMATION

It has been mentioned that some municipal parking authorities, such as Washington DC's Department of Transportation (DDOT), require by policy that, similar to coin payment practice, users must pre-commit to a certain amount of parking time at the start of a session. Entering the session time requires a few additional user interactions on the smartphone or vehicular user interface. To avoid this, a payment system could offer its subscribers the option of automatic renewal of the parking session in small quantities, such as five minutes. The user would be expected to close the session before leaving the parked spot. However, if he forgets, it is technically feasible to automatically sense movement of the vehicle away from the parked spot and to close the session with a message to the user, as described.

SESSION CLOSING AUTOMATION

There are benefits for enforcement process to automatically determine if a vehicle has moved away from its parked location. For example, if it is determined that the vehicle is still parked at the same location and the paid-for session has expired (in spite of reminders sent to the user), it would be known with certainty that the vehicle is in an *expired-session* state only if session closure was also guaranteed. What a payment system could do with this information is to be understood. Details could be found in TWS's patent filing.

NEW EXPERIENCE FOR ENFORCEMENT OFFICER (EO)

There may be one notch improvement of parking enforcement during the process of mobile payment. The objective is to eliminate it fully under the banner of RWPTTP when auto identification and auto invoicing are introduced. Enforcement is a major problem in most major cities owing largely to manpower shortage. In spite of under-enforcement, it has been reported that the fine revenues of major North American cities exceed the parking revenue by over three times.[7] Unlike private garages, city parking authorities, by charter, are not *for-profit* enterprises. Not surprisingly, their public position is that they accord priority to the first outcome. However, regardless of priority, it is clear that if information about the location of vehicles in expired-session states were made known to the EO, it would greatly lighten his/her workload. This would be desirable to the city. It may be noted that legacy pay-by-cell systems (which cannot determine the session status of the vehicle) have actually increased the Enforcement Officer's workload. To upgrade the scenario of off-street parking under garage with auto-mated vehicle and geo-stationary drone technology for auto identification and auto invoicing, a very high precision geolocation technique will be required.

PAYMENT SYSTEMS FOR GATED GARAGES

Pay-by-cell for parking in gated garages is far less common than for street parking but is beginning to emerge, at least as a marketing gimmick. In most cases, these systems use parking meters *inside* the garage, making them, effectively, an indoor extension of on-street parking. Unless the access to the garage is controlled and linked to the

---

7 In New York, in 2010, the fee revenue from 99,843 on-street parking meters was $140m and from fines $575m. www.bloomberg.com/news/2011-02-25/new-york-city-asks-bankers-to-plumb-parking-meters-garages-to-add-revenue.html.

payment system, we do not refer to it as a 'gated garage payment system'. In that sense, pay-by-cell for gate garages is still at a very nascent stage but needs to be explored for a value proposition. Then, what could be the value proposition of a *new* infrastructure? The value proposition of the new infrastructure that should last for another 100 years is in (a) taking advantage of available technologies and exploring on-the-horizon technologies, (b) establishing a next decade automated transportation paradigm, (c) the cost reduction, (d) establishing a gold standard of user experience, and (e) a smooth transition from the old to the new – a comfort for all stakeholders.

EXITING WHEN TICKET IS LOST

In present garages using paper tickets, if a user loses his ticket, he has to pass through a manually controlled gate where he has to pay a maximum charge which is set at a penalizing level while covering the garage owner for likely actual charges. In some garages, the user can call a remote, human operator who will collect a credit card payment and open the exit gate. This is onerous for both the garage and the user. In the TWS system, as the user can identify himself by his handset to the Parking Application Server, his ticket can also be identified in the Garage Server's database. He can therefore be charged the actual amount corresponding to his particular session plus a small penalty set by garage policy, designed to deter users from deliberately discarding their tickets. The lost ticket exit method involves a slightly longer transaction delay at the exit gate than if the user had the ticket. The garage owners may therefore wish to discourage users from defaulting to this method too often – leaving a scope of total automation under the RWPTTP banner.

PAYMENT AT MANUALLY OPERATED GATED GARAGE (VALET PARKING)

Many gated garages are manually controlled. This means that human operators, not computers, control the gates and/or collect payment through manually swiped credit cards. This is often the case with small garages in dense urban areas where small lots have been turned into paid garages. Manual operation is also the practice in valet parked garages, which are common in very densely populated cities like New York. The TWS system supports a cellular-data payment method for such scenarios, similar to mobile person-to-person payment. In this document, this payment method is referred to as 'direct payment' as shorthand for direct payment by the user to the garage (the payment amount is user-entered as opposed to being server-determined). Mmany benefits are seen by TWS system over the current system but automation of RWPTTP can supersede them all once each item of TWS is correctly identified and considered.

BROADCASTING BULLETIN BOARDS TO PROSPECTIVE CUSTOMERS

A garage parking payment system, such as the ones described above, could broadcast bulletin board information to proximate users. This is enabled by the following aspects of the systems: (i) the Garage Server is connected (e.g. via the Internet) to the Parking Service Provider's Parking Application Server, (ii) the Parking Application Server becomes aware of the locations of the smartphones when they initiate a parking transaction (indicating a desire to park at a gated garage), or are simply looking for a parking garage and request information from the server of available garages in a neighborhood. The bulletin board information may be updated as frequently as desired by the garage, another genesis of RWPTTP for PARQUEST features nationwide. The bulletin board of a garage can be pushed out to a user *before* he enters the garage, as soon as he selects <Garage Parking>. When the user selects <Garage Parking>, a message is sent to the Parking Application Server indicating the user's present location. In response, the Parking Application Server sends the smartphone the locations of proximate garages with which the Parking Service Provider has business relations. The locations are displayed on a geographic map. The user can view the bulletin board of each garage by selecting the corresponding green dot on the smartphone screen. The bulletin board may show the current parking rate, occupancy levels, future promotions, and any other information the garage owner may wish to advertise. The information may persuade the user to pick one garage over another. Hopefully, the RWPTTP infrastructure can extend the feature nationwide with the much friendlier user interface of the smartphone.

NEW TECHNOLOGIES: STEPPING STONE TO A NEW INFRASTRUCTURE

Emerging technologies that can benefit the parking industry are being reviewed here. Fortunately, the R&D for much of this development is being carried by other applications with huge scale, such as the Internet of Things (IOT) and high precision satellite navigation. Technologies that can aid parking services include the following:

- Connectivity technologies (wide and local area);
- Geolocation technologies
- Satellite navigation (GPS and other constellations)
  - Inertial navigation (ranging from simple wheel motion sensors to complex IMU subsystems)
  - Short range beacons (RFID and Bluetooth)
  - Proximity sensors (Near Field Communications, or NFC; infra-red sensors, optical pattern scanning).

CONNECTIVITY TECHNOLOGIES

At the core of a new wireless parking system is a wireless device connected to the parking application server on the internet. The application server, in turn, may be connected to many databases, through which identification, availability, rate for charging, and other enforcement procedures are made to a user account.

CRITICALITY OF GEOLOCATION TO PARKING CONFIGURATION AND PAYMENT SYSTEMS

GPS[8]-based vehicle geolocation and also geo-stationary or flying drone (regulated) will play an important role in parking configuration and payment systems and later self-identification, and self-invoicing as part of RWPTTP implementation, although it has not been widely exploited to date.

---

8 Smartphone vendors are increasingly using the term, GNSS (Global Navigation Satellite Systems) instead of GPS as modern smartphones can track additional satellite constellations beyond GPS. Today, both iPhones and Android-based phones are capable of tracking both GPS and the Russian constellation, GLONASS. It is expected that, in the future, the European Galileo constellation and the Chinese Beidou constellation will be added to the capabilities of high end smartphones, as well as tracking multiple frequencies (L1, L2 and/or L5), which helps to improve accuracy. In some markets, such as India and the Far East, tracking of regional constellations, such as GAGAN and QZSS respectively, may also be supported. Tracking multiple constellations and frequencies are expected to substantially improve the accuracy and reliability of the positioning system in the phone.

VEHICLES AND SMARTPHONES WITH GPS IN FUTURE

We need a standard and unassisted GPS, referred to as the SPS (Standard Positioning Service) by UASF (which owns and operates GPS), with high precision. Typically, the current standard provides an error of 2–10 m (sometimes in tens of meters in adverse conditions). Modern GPS receivers in smartphones are able to provide better accuracy by using the Wide Area Augmentation Service (WAAS) provided in-band relative to the GPS L1 frequency by USAF and, therefore, requiring no changes to the GPS receiver hardware. When assisted by WAAS, GPS accuracy is typically better than 1 m. Advances in satellite navigation have been made over the past decade for applications such as precision farming, mining, and surveying that enable accuracies to centimeter levels to be realized at a high costs, but at a very limited scale levels. This class of advanced satellite navigation technologies is collectively referred to as High Precision GNSS. While it is still unclear to what extent smartphones and vehicles will embrace these technologies (hoping to drive cost figures down by several decimal points) it is clear that driver-assisted/driverless cars of the future will require these technologies – they become the main scale driver in the future. Furthermore, these satellite navigation technologies will need to be sensor-fused with other navigation technologies such as LIDAR and inertial navigation to create integrated navigation systems with greater accuracy and robustness than any of the individual components of the integrated system. Such advanced satellite navigation systems have immediate applicability to on-street parking and later on garage parking. Today many on-street parking payment systems are based on zone IDs (posted at the parking locations) which the user has to enter on his/her smartphone. Geolocation with decimeter to centimeter accuracy makes zone IDs unnecessary later when parking application can monitor the location of the vehicle on real time in the parking location.

OTHER NON-SATELLITE LOCATION TECHNOLOGIES

In addition to satellite-based navigation, geolocation may also be performed by smartphones based on RF beacons. These are proximity-based systems where the location of the access point or beacon transceiver is known, and the RF coverage is sufficiently small that the location of access point suffices as a proxy for that of the smartphone. The RF technologies include WiFi, Bluetooth, Near Field Communication (NFC), and infrared

(IR). In this category, one may also include optical scanning of a pattern that indicates location. The beacon may use an IFF transponder-like protocol. In such protocols, the beacon will periodically issue a query with its own ID (I am John, do you hear me?); the responding device will respond with a response and include its own ID in the response (I am Jane, I hear you). This informs the two devices of each other's ID and that they are within coverage range. One example is as follows. Assume that a Bluetooth Low Energy (BLE) beacon is located near a garage entrance gate, announcing the garage/gate ID every second. A user in a vehicle, carrying a smartphone with BLE support, drives up to the gate. The smartphone becomes aware of the garage/gate ID and communicates this to the Payment Application Server. The Payment Application Server determines the IP address of the Garage Server from the ID of the BLE beacon and opens a communication link to the Garage Server, wherein he requests the latter to open the relevant gate.

Modern smartphones are able to scan 2D patterns such as QR codes[9] using the phone's camera and associated software. Therefore, these codes can be used to provide the same information as the BLE beacon described above. The same is true of NFC, where a very low power beacon is provided on the phone and the access point is part of the infrastructure. In both of the above cases, the phone can learn the ID of the access point. However, unlike the BLE case, a greater level of user action is required. In the BLE case, a use case involving *no user action* (the ideal user experience) may be realized by suitably addressing some 'corner cases', such as the possibility of interference from adjacent beacons.[10] Even if *some* user action is required, it is limited to two to three touches on a phone screen, which is preferable to requiring the user to wind down the window and bring the phone close to an access point. In any case, there appears to be little benefit to using a phone over a credit card (beyond the ability to store the card in the phone) if the user has to handle both in the same way. Currently available and evolving technologies may be packaged in various forms to create different user experiences.

---

9 *QR code* is the trademark for a type of matrix barcode (or two-dimensional barcode) first designed for the automotive industry in Japan (Wikipedia).
10 Technologies exist to address potential inter-beacon interference. One is called 'fingerprinting' and involves recognizing a pattern of signal strengths from a number of BLE beacons that surround the location of interest.

ADVANTAGES OF INTEGRATED ELECTRONICS IN THE VEHICLE

With the level of integrated electronics in vehicles expected to rise dramatically to support future driver assistance and driverless cars, the parking payment solutions described above are all likely to be free byproducts of the above electronics. Primary enablers for parking applications are expected to be Bluetooth, perhaps RFID, integrated data connectivity (cellular and WiFi), and high precision navigation (involving a fusion of assisted GNSS and inertial/optical sensors).

When these features become available, parking payment systems including other transportation services should be able to deliver the ultimate goal of *complete user transparency* described above.

DATABASE CONFIGURATION

The two systems (SMA™ and TWS) discussed above and other systems referred in this chapter reflect one common theme – gathering relevant information in the form of a limited data types and use of it for a solution mostly locally. In the grand scale of future national evolution with RWPTTP, they may be termed as databases: database of vehicles, database of drivers, database of parking spots, database of rates, database of available spots, database of garages, database of traffic congestion, etc. Moving to autonomous and electric vehicles, there are other data types that are collected and may be very useful for a vision of autonomous transportation systems including no meter and no gate, just auto identification and auto invoicing. Below is a preview of those data and their proper use for the benefit of all stake holders.[11] Parking industries are fragmented. As a result, data collection of parking services is also fragmented. Cities or municipalities collect data at some levels with the help of parking operators. These data can hardly be used for consolidated or centralized planning of smart cities. Cities or municipalities still use long outdated empirical formula for planning parking facilities. So far, data generation, and collection or use of it in the parking industry, has not been streamlined. The above statement may not apply completely for transportation services, especially for intelligent transportation services (ITS). ITS started collecting some data

---

11 Amalendu Chatterjee, 'Autonomous Vehicles, APPs and Integrated Transportation Services', *Parking Today*, February 2018.

several years back under the sponsorship of both Federal and State Department of Transportations (DOTs). In parallel, DOTs have developed well managed regulations for the proper collection (mostly manual) and use of those data such as number of traffic tickets issued, speed limit violations, etc. Electric and autonomous vehicles have added another dimension to their initial plan. In addition, those days may not be very far off when autonomous vehicles (self-driving or self-parking), smartphones, their applications, and integration of parking and intelligent transportation services may become a normal part of life. It is time to explore different sources of upcoming transportation and parking data, owners of those data, regulations surrounding the use of those data, and benefits of those data for the stakeholders. Those important areas of parking need to be studied thoroughly for a better understanding and potential future data analysis in the face of connected devices such as IOT and cloud computing – use of online learning and decision making. The IOT phenomena enhanced with AI will further enhance the scope of data generation and collection of data – the extent is yet to be ascertained. Such analysis of combined or integrated parking and transportation services is required for devising needed new regulations for the security, safety, privacy, efficiency, and convenience of all those involved, such as customers, product manufacturers, implementers, enforcers, and others. You can imagine the range and scope of data when Figure 4.1 is fully implemented for the future of transportation services while various devices starting from parking lots, trucks to kiosks, drones, etc. are interconnected in the web platform.

DATA SOURCES

You are familiar with the concept of a black box in a plane. It records all conversational and navigational data in the flight. There are two components of the black box – Flight Data Recorder (FDR) and Cockpit Voice Recorder (CVR). Both are required by Federal Aviation Authority (FAA) regulations for commercial, private, or corporate aircraft. CVR records radio transmissions and other strange sounds in the cockpit. FDR records altitude, airspeed, and direction of the flight. The black box is located in crash survivable part of the aircraft and protected from water damage with a device called Underwater Locator Beacon (ULB), the 'Pinger'. Data gathered in the black box are used to analyze reasons of the crash – pilot errors, mechanical errors, weather conditions (too cold or thunder storm), or something else. Such analysis may possibly improve the safety of future aviation as technologies are newly discovered.

Imitating the concept of CVR and FDR, all autonomous vehicles are installing similar black boxes but with different capabilities appropriate for integrated transportation services. Sophistication of such a black box in vehicles are not required by strict government regulations yet. But vehicle manufacturers are aware of such benefits as they are making inroads with self-driving, self-parking, and other autonomous functions and features for the convenience of riders. Smartphone applications by Internet companies and online exchanges while driving used in the transportation services will also generate much more data than previously envisaged. Shaping the regulations, taking account of different data, is important but more important is the scalability and data extents or usability. BDA or BDM tools available in the marketplace for other industries need to be explored to define final criteria of regulations. Areas of such raw data:

- Phone conversations with built-in cellular connections
- Text exchanges
- Speed
- Seat belt status
- Baby seat status
- Temperature of the vehicle inside during extreme weather
- Queries to websites while driving
- Music tuned
- Panorama view of the surrounding area while driving
- Driving habit – cruise, speed, etc.
- Parking habit – on-street, off-street, etc.
- Parking location
- Office commute habits
- Shopping/buying habits
- Collecting web browsing history

- Accident or incident on the way in or out

- Visit to gas or charging station

- Weight of the driver or the passenger

- Number of occupancies

- Medical conditions of vehicle passengers

- Alcohol level

- Eye movements while driving and driver's sleeping record

- Hands on/off the steering wheel

- Tire pressure

- Vehicle performance based on design criteria

- Miles driven for each trip

- Number of passengers and reasons of the trip

- Info on loads in a truck, content, destination, driver identity

- Miles driven before recharge/refueling

- Congested area for alternate route

They are not exhaustive. Other data are generated while you activate different apps while driving or riding autonomous vehicles. Figure 4.2 is a snapshot of my Tesla model X (autopilot platform) touch screen with different icons. A click on any icon generates data of specific data type and is subject to recording either in the vehicle black box or in the specific company website.

The above phenomenon is true even when you share a car ride or public transportations. Implications of these data may lead us to different set of regulations once we understand them well. There may be some commonalities of data with the transportation services. But transportation services are much wider and are managed and regulated by different federal and state government agencies under the broader umbrella of department of transportation (DOT). Federal agencies are: federal aviation administration (FAA), federal railway administration (FRA), office of the secretary of transportation (OST), federal transit administration (FTA), federal highway administration (FHWA), federal motor carrier safety administration

(FMCSA) and national highway traffic safety administration (NHTSA). There are parallel or similar state transportation administration agencies, but they differ state by state to complement federal agencies. Agencies that are relevant for this chapter are related to national highway transportation, trucking and motor coaches, automobiles, and public transit. These agencies collect data, regulate industries, and provide updates to regulations as and when necessary. Relevant data for their functions are:

- Arterial and highway vehicle volumes

- Vehicle classification counts

- Intersection turn movements

- Auto occupancy counts

- Interstate corridor travel time and delay

- Pedestrian/bicycle counts

- Origin and destination hubs

- Video license plates

- Parking utilization

- Radar interception counts

- Traffic violations

- Speed of vehicles and exceeding range limits

- Erratic driving data

- Blood alcohol contents (BAC)

- Drunk driving data

- Fatalities due to driving while intoxicated (DWIs)

DATA COLLECTION STUDIES FOR FRAMING CORRECT REGULATIONS

Technology capabilities for generating data and its collections are limitless for services, especially when autonomous vehicles are connected to IOT. The objective of these data collections and surrounding regulations are to set limits and relevance, including ownership, while providing new transportation services. Both over-regulation and under-regulation must be avoided. Some data are useful for city planning, reducing congestion, automating traffic management, and supporting 'Go Green' to help the planet. Some may be useful to the driver while driving, some may be useful for the enforcement officers, some may be helpful to new entrepreneurs, some may be useful for generating new revenues, and some may be useful for market promotion and advertisement. Other industries (FAA, FRA, etc.) have been collecting similar data and they were found useful. But their extents were well defined, limited, and well bounded with appropriate responsible ownership – mostly government. Autonomous transportation services have shifted the balance. To serve our industry better, our discussions may best be guided by the following yet to be refined outlines to set the boundary for all.

Who Owns the Data and What Can Be Collected

There are many players in the game plan with differing interests. Ownerships, boundaries, and the authorities are to be clearly defined to preserve the interest for fair market dynamics. Hacking of these data must also be protected to prevent ransom hold ups. Autonomous and remote operations of vehicles on a continuous basis are also subject to hacking of keys and codes for cyber burglaries. This is the opportunity to define the concept of a black box for autonomous vehicles so that data can be shared during crash investigations.

What Should Be the Regulation behind Such Data Collection?

Regulations may be defined in many ways. Regulations requiring boundaries, partitioning, or openness before data generation and collection are initiated. New regulations are required to control and manage stored data ownership. Updating older regulations in the wake of modern technologies will also be helpful for a new transportation platform and configuration.

CAN DATA COLLECTIONS BE COMPLETELY AUTOMATED

There may be different options: mechanical, electronic, bio-tech, or a combination of the above. Ignition Interlock Device (IID) or Breath Alcohol Ignition Interlock Device (BAIID) to prevent drunk driving is one such device available in the market. The device is not foolproof yet. Such early devices had semiconductor-based alcohol sensors and did not hold calibration well. In addition, they were sensitive to altitude variation and reacted positively to non-alcoholic sources. In 1990, the industry perfected the device with fuel cell sensors. The idea behind IID or BAIID is to interrupt the ignition signals to the starter until the ignition is satisfied with the alcohol levels from the breath sample, according to the state the car is registered in. To prevent cheating by someone else, the driver needs to pass another test at random several minutes after the start. The field is wide open for new innovations. Bio-tech sensors may be incorporated in the autonomous vehicle for drunk driving, similar to the digital medical pill embedded with a wearable sensor to signal the doctor remotely.[12] Data collection can be automated and also analysis of it can be instantaneous.

WHAT ARE THE BENEFITS/HARMS OF SHARING SUCH DATA?

Actionable insights of all these data are critical. BDA tools must be deployed to visualize data in seconds with ease of use. Storing data in the cloud platform will assist for automatic update and sharing with the right individual. Many benefits could be derived, with various data related to convenience of riding, auto invoicing, autonomous payment, selection of best or alternate route to destination. Marketing of data collected will create an economic boom in the industry. Along with it will come the fight for ownership and litigation. If data are compromised due to hacking, privacy could be an issue – loss of identity. There are many benefits of collecting all these diverse data, but a good strategy must be thought through.

A Future Parking Paradigm: No Meter and No Gate but Auto Identification and Auto Invoicing like Automated Toll Services ('no Stop & Go')

With the latest advent of technology, the parking industry can change vastly for the better Automation (auto identification and auto invoicing) will lead the way. Once the license plate is captured with high speed camera accuracy over 99.99%, a database look up of the national DMV can identify the vehicle owner to send the invoice to his address if the RFID tag is not registered with the transportation authority. If we can bring the IOT compatible drone technology to auto monitor the traffic

---

12 Pam Belluck, 'First Digital Pill Approved to Worries about Biomedical "Big Brother"', *NYT*, November 13, 2017.

on the interstate highways, we come closer to an integrated revolutionary paradigm – a solution much better than recommended by Prof.
D. Shoup and his colleagues in *Parking and the City* of eliminating off-
street parking (otherwise known as PARCS) in favor of on-street
parking with fair prices. In my opinion, their solution does not go far
enough in addressing the real problems of overbuilt off-street spaces in
city business centers, inefficient utilization. We need to move spaces to
suburban areas, including having a moratorium of off-street spaces
(eventually) to align with autonomous transportation, and technology
that may cover many existing issues and improve costs nationwide.
Parking spaces are national resources that require private and public
consolidation and must be raised to that level for a standardized uniform solution. A national database is required for equitable distribution
in a convenient location. Such a database, along with autonomous
vehicle database, iCloud platform, BDA, AI, no meter and no gate,
sophisticated software instead of field hardware, and personal devices
such as smartphone with smart apps, will assist autonomous identification and autonomous invoicing implementation, dictating the future
transportation paradigm. One can get a preview of such paradigm
through SMA™ and TWS's mobile payment system described in the
sections above. Customer-related administrative responsibilities could
be shifted to end users following the model of airline passengers. End
users will do most of the work to get their wish granted – a one-click
solution. The parking solution should be the tail end (last mile) of other
integrated virtual transportation services. I contend that the evolved
process (no meter and no gate) I propose via RWPTTP, hopefully, will
support Prof. Shoup's third theory, 'Spend the Parking Revenue to
Improve Public Services on the Metered Street', with more revenues
from new services. Other supporting arguments are given below.

An Innovations and technologies have revolutionized our social
fabric over the years disrupting established industries.. Millennials' car
ownership and driving habits will be different from those in the last
century. Even the Department of Motor Vehicles (DMV) may go
paperless with electronic license plate displays. The parking industry
must be prepared to embrace radical change, rather than incremental
adjustment. In addition, strict enforcements give an unfriendly image of
the industry – so why not let the end user do their own enforcement?
Mobile and credit card payments have been added for incremental
values but not at the level consolidated technologies can provide. The

upgrades needed include a national parking database for proper counting of resources, private and public parking space coordination, integration with other transportation services such as train, airline, hotel, car rental, etc., synchronization of parking services with other virtual transportation services such as Uber/Lyft, and driverless transportation services. The majority of spaces in the current configuration are in open lots such as airports, shopping centers, hospitals, commercial buildings, university campuses, city business centers, etc. providing flexibility for new innovations. More values will be added when garage parking spaces are eliminated with relocation if needed. In this context, truck parking should also be contemplated. Exploring common technologies for the scale of economy is the main theme of RWPTTP with different databases nationwide. Such database formats require a national and international standard for a uniform platform or infrastructure for information sharing. That is why the International Parking and Mobility Institute is collaborating with European parking communities for a similar vision. A new approach does not need to differentiate between on-street and off-street parking. In fact, the concept of garage parking in the city area can be totally eliminated requiring to define a suburban parking district (SPD) and move all on-street and off-street parking spaces from the city business center (CBC) to SPD so that people can walk freely with no car cruising congesting the city street. Automated transport vehicles should transport people from these SPDs to the city center. Meters and gates are not required and the city land can be restored for urban housing and other economic development to minimize rural and urban economic gaps. All parking spaces will be available for online reservation remotely anytime and from anywhere for a one-click solution – similar to locating a supercharging station for Tesla EV. Similarly, one can define a new Truck Parking District (TPD) like SPD in the Interstate Highway corridors in coordination with weighing stations – a much needed requirements. Such approach will enhance the transportation of goods. Imagine the economic growth expected if both SPDs and TPDs are combined under one infrastructure of a robust transportation system with common but standard technologies which will also open the opportunity to promote renewable energy.

Ideas expressed here are not totally uncommon in the context of the Internet or virtual infrastructure interconnecting devices using an iCloud platform – software emulation of hardware functions for many smart decisions. The iCloud platform is being enhanced everyday with AI to make statistical decisions. Wireless for mobility and data analytics (big data mining) for accuracy (data generated by all connected devices as required by regulations) will form the backbone. It will be like a concert where all instruments play in harmony. Here in the transportation concert, diverse technologies will play in harmony to connect or disconnect devices as and when needed. Geo positioning system (GPS), Bluetooth beacon, BLE, PV cells, EVs, drones, cars, trucks, etc. playing in harmony, connected via IOT for transporting people and goods.

High-level design objectives to be met are: configuration and reconfiguration of parking spaces to increase revenues at short notice, synchronous transportation and parking services, consolidated private and public parking spaces for efficient land use, self-identification of drivers and vehicles for auto invoicing, auto identification of valet and handicapped drivers bringing uniformity. In addition, one must explore covering every 500 open spaces with PVs to generate electricity (2.5 MW) for charging EVs and other use – self-sustainability. Hopefully, the process will reduce air pollution, simplify revenue collecting hardware at costs less than $40 per space compared to $800 per space now, open the possibility of a one-click solution. In fact, the payoff period of new technologies may be shorter than any other existing business model.

NETWORKING ARCHITECTURE FOR RWPTTP

RWPTTP will need a strong global communications network connecting many other subordinate networks of remote sites to solve transportation related services. Such networks or combinations of networks will have many remote sites, will grow, and will have many devices (roaming or stationary). In other words, such a network will be characterized by on-demand adjustments of physical connections (wireless and fiber), virtual configurations within the physical connection, and data collection for constant updates.

Volumes of papers and books have been written on the network architecture since the Internet and different social media became the part of our everyday life. The RWPTTP architecture may not be much different. Obviously, expert help from those areas may be borrowed. There is some obvious difference in two areas: billions of devices connected to RWPTTP networks are of different types with diverse functions. In addition, a revenue sharing mechanism among its partners connected to RWPTTP (parking owners – public and private, parking operators, other stake-holders) needs to be embedded. Every partner will interact at different levels to develop applications to satisfy end users. The network architecture of RWPTTP may be similar to a social media network such as Google in some aspects, but may be similar to the telecommunications network for revenue sharing. Irrespective of these differences, the network architecture must consist of robust elements of the following.

- Physical Network Architecture also called Next Generation Net-work (NGN)

- Virtual Network Architecture also called Software Defined Net-work (SDN)

- Database Network Architecture also called Data Centric Net-work (DCN)

Physical Networking Architecture or NGN

It will be a combination of edge and backbone. Basically, it will define physical elements of hardware devices forming a local or regional network. Wireless network (such as 5G) components will play a major role in both the access and the backbone. That is why physical network architecture must be robust and scalable. Because, we are talking of about billions or trillion devices physically connected together to provide the real-time service as and when required and wherever required. That is why NGN concepts are embedded as built-in as architectural changes in the core and the access. The purpose is to transport voice, data, video, and multi-media, separating the transport (connectivity) portion of the network and the services part that run on the top of the transport. Such a capability enables a network provider to define a new service without disturbing the transport layer. NGN though built around the Internet is more than just the future Internet evolution.[13] We know, for sure, NGN may not be called acircuit switch based telephone-centric network because it is built on Internet Protocol (all IP – end to end). ITU-T defines NGN as packet switched network to provide a broadband quality of service delivery independent from underlying transport-related technologies. NGN offers unrestricted access by users with generalized mobility which will allow consistent and

---

13 Wikipedia.

ubiquitous provision of services.[14] Naturally, NGN will need three main core architectural changes: (1) public switched telephone network (PSTN) to voice over Internet Protocol (VOIP) – a capability of emigration and emulation of legacy services, (2) changes of switching infrastructure, and (3) migration of constant bit rate service to a variable rate packet service with optimized latency.

Virtual Networking Architecture or SDN

As known today, the future network will be more software defined and managed. A physical infrastructure is configured for the extent of its physical reach. SDN can make the network presence basically virtual, enabling flexible and agile operating system interactions irrespective of its physical layers. Cloud computing is one such example. 'The Virtual Cloud Network delivers secure, pervasive, connectivity for apps and data, wherever they live – from the center to the cloud to the network edge' according to some experts of VMware NSX technology.[15] SDN enables cloud computing and network engineers and administrators to respond quickly to changing business requirements via a centralized control console. SDN can support multiple kinds of network technologies designed to make the network to integrate the virtualized server and storage infrastructure of the modern data center. Software-defined networking has perfected an approach to designing, building, and managing networks that separates the network's control or SDN network policy (brains) and forwarding (muscle) planes, thus enabling the network control to become directly programmable and the underlying infrastructure to be abstracted for applications and network services for applications as cloud computing or mobility. All seven layers of open system interconnections (OSI) model of a full-blown (end to end) digital network are used to implement SDN.

Database Networking Architecture or DCN

There will be volumes of new data with different characteristics not seen so far for the full implementation of RWPTTP. Real-time interactions in the integrated transportation systems will be a norm. Such operations with new types of data will need an efficient database design.[16] All data

---

14 Tsbedh. 'NGN Working Definition' (www.itu.int/ITU-studygroups/com13/ngn2004/working_de finition.html).

15 Ken Kepes, 'Rackspace Support Network Report' (http://diversity.net.nz).

16 Wikipedia, December 10, 2018 update.

types may not be useful for new application developments. So it is important to follow certainsteps while defining new and old database design, because blending the two will be necessary in some cases.

Defining Data Types to be Stored

In the face of automated transportation, many data types will be generated. Initially, their use may not be known but all need to be stored. Experts with domain knowledge will transform these raw data to useful formats with proper requirement analysis. Such requirement analysis is done according to the standard specification. These data can then be stored for defining applications and services as required.

Identifying Relationships among All Data Types

Here we are talking about old data, new data, different domain data (parking, traffic, toll, DMV, vehicle, etc.). Dependency among these data must be established. Why? Because changes in one may trigger changes in the others – without you knowing it in the absence of this proper relationship. This is called data relationship and could be used for innovative service definitions. There are two kinds of relationship: dependence relationship by data elements and entity relationship by attributes. Such relationships could also be used for the security and the safety of the transportation system and the public. System designers must be aware of data that have been stored, to think of their use as and when they are to be activated; be it a new service definition or deriving other new data elements.

Logical Structuring of Data Types

First, one has to determine the purpose of each data type. Then, organize the information in the data in a logical format (maybe a table) keeping in mind that data relationship may be simple (child–parent) or complex (child to multiple parents). Each table will follow a standard row column structure. Logical structures are required so that these data are amenable to a standard database management system (DBMS).

Normalizing Necessary Data Types

Normalizing data types is required for designing a system that uses such data. Normalizing means to deal data in a systematic way when querying for some information without losing any data integrity. Normalizing is also required when data need to be updated. The database design also uses hardware, software, and storage media for high performance that is essential during retrieval process.

USER PERSPECTIVES

From user perspectives, a worldwide web (WWW) view for all transportation services is required. We can think of defining a DNS as 'RWPTTP' by name. Users will click www.rwpttp.com for all their transportation needs for convenience and new experience. The first phase of RWPTTP is 'No Meter and No Gate: Auto ID like Automated Toll Services' in Interstate I-540 in North Carolina. Auto ID and auto invoicing will bring at least toll and parking services together. Similarly, auto traffic management using drones (for speed and irregular driving) can also be integrated to a common view of all vehicles. Slowly, we will enter into a paradigm of vehicle management that provides transportation services – be it cars, trucks, buses, etc., or be it roads or in parking lots using national and local road infrastructures. I call this national resources. That would mean many new user-friendly regulations with minimum human interactions, reducing infrastructure costs. It would also mean users take control of their required activities to manage their accounts and services. With such an approach, the strategy is to minimize fragmented infrastructure, enabling centralized data collections of parking services aligning with other transportation services. Many old regulations are unnecessary and need to be phased out in this new paradigm – a paradigm of national standard for uniformity. New regulations for a paradigm change have to be socialized for the benefit of socioeconomic development. Currently, cities or municipalities collect data at limited levels with the help of parking operators with no repository for global use for consolidated or centralized planning of smart cities. Cities or municipalities still use long outdated empirical formula for planning parking facilities. So far, data generation and collection or use of them in the parking industry have not been streamlined. This statement may not apply completely for transportation services especially for ITS, which started collecting data under the sponsorship of both Federal and State Department of Transportations (DOTs). In parallel, DOTs have developed well managed regulations for the proper collection and use of those data. Electric and autonomous vehicles have added another dimension to their initial plan. It is time to explore different sources of transportation and parking data, owners of those data, regulations surrounding their use and the benefits for stakeholders. Such analysis of combined or integrated parking and transportation services is required to devise new regulations for the security, safety, privacy, efficiency, and convenience of all those involved.[17]

---

17 'Parking in Perspective The Size and Scope of Parking in America', National Parking Association Report, May 2011. http://en.wikipedia.org/wiki/E-ZPass.http://us.parkmobile.com/pdf/83330aa4dc.pdf.

CHAPTER 5

Single Payment Method

INTRODUCTION

The subject of parking payment options and their charges or fees to the parking operator and, in turn, their customers, needs new investigation. The objective is to explore if such fees can be reduced or even eliminated so that arbitrary parking rate increases or decreases can be rationalized to consumers. The other objective is to avoid subsidization of airport/city's other budgets from parking revenues. Payment processing for the parking industry started almost 100 years ago. The mode of payments was cash only during the early stage of the parking system. The history was well presented by Mike Drow of SP+ and Peter Lange of Texas A&M University in the 2014 IPMI Conference. The on-street meter capability for accepting coins started in 1935. Off-street acceptance of credit cards started in 1950 with cashiers. Automated credit card access and exit started much later. Options are so many and still evolving, Expected Monetary Value (EMV) chip and blockchain (with Bitcoin being subset) are the latest. The question is what to choose and how or why to choose. For large or medium size parking owners or operators, all options may be required to satisfy the variety of customers under the current paradigm of the parking business. In my presentation at the same conference, 'The Zero Cost of Civilized Parking', a new paradigm with Robust Web Parking, Truck & Transportation Protocol (RWPTTP) was defined. I also called it Affordable Parking Act (APC). In that paper, I showed the cost of on-street and off-street configurations could be reduced drastically with a new paradigm. In this chapter, some methods will be highlighted to reduce the cost of parking payment options and their processing fees. RWPTTP, as a one-click solution for parking, may accelerate the process when blockchain payments are regularized and become part of it.

PAYMENT CHOICES

Choices and options of payments evolved for relative convenience to the customer. But what is the ultimate convenience for all stakeholders? – that is what we should shoot for. There are rationales for many step-by-step improvements of choices. These choices also improved the security of payment systems and decreased fraudulent activities. This latest of all choices is EMV, providing all benefits, especially the security of payments. This, in turn, raised the costs for the parking operator and the owner, even if we ignore the latency of processing. It is time to go back to basics so that choices are reduced but we still achieve the flexibility of modern payments and the required security.

Comparing Payment Systems of Other Services

For wireless phone, electricity, city water service, waste collection service, and other services each has always one simple option for payment collection – invoice or bill and then the payment by cash or check or credit card by phone. Is it possible to adopt one unique system rather then so many options for parking payment? Let us investigate that theory. For analysis, we can divide parking payments into three major groups:

- Traditional – Meter, Pay-on-Foot, Pay-In-Lane equipment, Event Handheld, Cashier/Fee Computer, Credit Card/EMV, Website, Kiosk, Mobile Payment (Square D, PayPal)

Days of these options may be counted. They are expensive in two ways – setup cost, maintenance, and processing fees.

- E- Wallet – Google Wallet (MasterCard), Apple Wallet, V.me (Visa)

This may be an interim solution before virtual currency takes over. The advantage of E-Wallet is a consolidation of too many cards. In addition, attempts are made to reduce the processing fees.

- Mobile Payment by Cell

This is an end user initiated program. You subscribe to a parking application in a smartphone and then activate your parking option while in the meter space. It needs a couple of steps before the meter payment is accepted – not very customer friendly. TWS described in Chapter 3 has a patent to reduce such steps by using a BLE beacon in the vehicle and communicating with the smartphone with a pre-registration.

- Virtual Currency – Digital Wallet Plus (DW+), Bitcoin (Digital Mining) & others

This option may be the ultimate disrupting technology of payments and financial transactions. If it is widely accepted in the industry, current processing fees may be reduced to almost nil.

Cost Comparison

Cash components of traditional parking payments are many, such as labor, counting, storage, collection, delivery to bank, cashier booth, processing fees, card terminal, PCI compliance, administrative cost, exception handling and refunds issued, hardware units, associated software updates, signage, and infrastructure. In a large airport, almost all options such as credit card (76% collection), cash (22% collection), and other (2% collection) are deployed and the total cost of payment processing is over 17% of total collections as per the paper presented at the 2014 IPI annual conference. The same total collection costs can be reduced from 17% to below 5% of total revenues if only the credit card option is deployed, eliminating cash and other options. This scenario of the cost optimization is applicable to the medium or large size city and university. The cost of new EMV cards (credit card with a chip not magnetic stripe) which is higher now have not been factored in here. We assume the mobile payment processing is similar to the credit card processing. But all are centralized control, lacking flexibility and convenience, with no customer responsibility.

If parking operators change the parking paradigm with a Web Parking Portal (WPP), customers will need to set up an individual account with a credit history for payment like paying for any other services. In this scenario, there is no need for credit card processing at site. Customers can pay (web account) online or from his/her bank account with no requirement of at-site payment processing. This may not be dynamic enough to keep with the pace of modern technology and the younger generation. That is why we have to evolve with time and technology.

Efforts are being made to reduce credit card processing costs further by introducing Google Wallet, V.me, and Apple Wallet. Their acceptance nationally and internationally is yet to be debated.

Is blockchain (and its subsets, Bitcoin and others) the cheapest and the latest method? Yes, if planned correctly. Bitcoin transactions are equivalent to cash transactions except that they are done via the Internet. You can call Bitcoins the cash of the digital virtual world. They have different denominations like coins or notes. These coins or notes are like digital logos of different sizes and colors also called Bitcoin Blocks. Bitcoin transactions (in contrast to credit card or bank transactions) are almost free except in some special cases and they are recognized internationally to a limited extent. Fees that are applicable are very small or flat (less than $0.4) irrespective of the amount of transactions as of today but may evolve to different fee structures (hopefully lower) as the market matures. Fees are paid to the Bitcoin miners (like credit card issuers) as their incentives for mining different Bitcoin blocks. With the digital Wallet Plus (DW+) and Virtual Currencies (Bitcoin), there is a very good potential for cost optimization.

TYPES OF CRYPTOCURRENCY AND ITS PREFERENCES

According to Wikipedia, blockchain is defined as blocks which are linked using cryptographic hash of the previous block, a timestamp, and transaction data generally represented as a merkle tree root hash. Figure 5.1 shows the blockchain formation. The main block consists of the longest series of blocks from the genesis to the current block. There are multiple types of blockchains or cryptocurrencies.[1] Notable ones are: public blockchains, private blockchains, and consortium blockchains.

Public Blockchain

Bitcoin and Ethereum fall in this category. With no public access restrictions to the network, anyone with Internet access can initiate transactions and become a validator. Public blockchain is a decentralized scheme

---

1 Wikipedia, 2018. Hyun Song Shin, 'Cryptocurrencies: Looking beyond the Hype' (www.youtube.com/watch?y=vo6s1mujxqq&list=pljkkw-wsobgqpz5mnl2sgmyyc78exhzrkh&index=2). Nicholas Weaver, 'Blockchains and Cryptocurrencies: Burn it with Fire' (http://youtube.com/watch?v=xchabodnnj4).

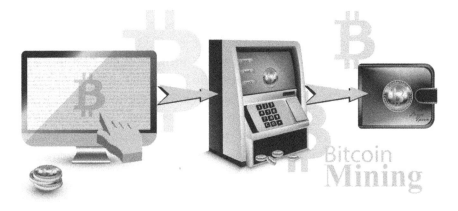

FIGURE 5.1   Blockchain formation.

avoiding centralized points of vulnerability. Every node system keeps a copy of the blockchain data replicating database with computational trust. Such schemes always require computer resources to process larger amounts of data.

Private Blockchain

Access is on a permission basis and someone has to be invited by the Internet administrator to be a validator. Private blockchain is preferred by private corporations who are not that comfortable with exposures of their private and sensitive information to the public network without any oversight or autonomy. Many experts think that private blockchains will be cumbersome databases and could be subject to suspicion.

Consortium Blockchain

This system works in a semi-autonomous mode. Each corporation can control one or two nodes of the network. The consortium blockchain also requires permission to access but is limited to corporations who control the node administration.

In spite of critical skepticism, many institutions and corporations are looking at its adoption to capture the market segment. Among them, banks are prominent. Others are sales organizations and organizations which manage online contracts. RWPTTP could also use it for its all transactions (auto identification, auto invoicing, and auto payment), based on its successful use by other institutions.

REGULATIONS FOR BITCOIN (CRYPTOCURRENCIES)

The State of New York is the first state to propose regulations for virtual currency companies operating within it (*NYT*, July 17, 2014). The regulation called 'BitLicense' plan, which includes rules on consumer protection, the prevention of money laundering, and cyber security, is the first proposal by a state to create guidelines specifically for virtual currency. The new rules, based on a comprehensive framework with flexibilities (based on technology) according to Benjamin M. Lawsky, financial regulator, New York, would be required for Bitcoin exchanges and for companies that secure, store, or maintain custody or control of

FIGURE 5.2   Bitcoin ATM machine.

the virtual currency on behalf of customers. Merchants that accept Bitcoin for payment (shown in Figure 5.2), like Overstock.com, would not need to apply for a license.

1) The rules for Bitcoin companies may be similar to existing regulations for banks and other financial institutions.

2) Bitcoin companies handling virtual currencies would be required to have robust cyber security programs against hackers.

3) As capital protection, but unlike banks, these companies subject to capital requirements would be allowed to hold some of it in virtual currency.

The objective of government in overseeing Bitcoin companies is to increase consumer confidence and to avoid wild price fluctuations. Hopefully, other states or the Federal Reserve will get involved to promote digital currency as an alternative to more advanced credit card payments that require middlemen processing.

Further research is going on at MIT and University of California, Berkeley, on the viability of the concept for wider adoption. In 2015, a new journal, *Ledger,* was announced to dedicate its contents on cryptocurrency and blockchain research. Submissions of papers and interactions with authors are managed by The Bitcoin blockchain.

CHAPTER 6

Parking Space as a Source of Renewable Energy

INTRODUCTION

Professor Varun Sivaram of Georgetown University wrote a book, *Taming the SUN: Solar is in my DNA*. According to him, it is not the tariff on Chinese solar panels but innovations to harness solar energy that will power the planet and every house, and also reduce environmental hazards.[1] The industrial revolution established different transportation network infrastructures for railways, airways, seaways, and roadways. These networks helped continuous economic growth and social evolution, promoting business opportunities since its inception. Over time, this transportation infrastructure has also become sophisticated but expensive. High energy usage is one of many reasons for high costs. Here I am talking of standalone Intelligent Transportation Services (ITS). Consolidating parking services with transportation services may increase additional energy requirements given the high space to car ratio. A number of over 900 million parking spaces for 300 million cars is floating in the US. On the other hand, there may not be enough parking spaces for trucks transporting goods door to door – a source of

1 *India Abroad*, September 14, 2018.

traffic congestion and GHG emissions during commuting hours. In the last estimate, there are over 15 million commercial trucks and only 308,920 truck parking spaces along interstate highway corridors (rest areas and private truck stops).[2] There are only 5,000 such truck stops in the U.S.A., not enough, causing trucks to park on access ramps (a traffic hazard). Each truck occupies close to three times one car space. Here I will be making a case that all these spaces will be candidates for a renewable energy platform for the future parking paradigm where each space will be either relocated or reorganized into a suburban parking district (SPD) or a suburban truck parking district (STPD) to be part of a web-enabled transportation platform (WETP), broadly called Robust Web Parking, Truck and Transportation Protocol (RWPTTP).

Background of Transportation Services

Private vehicles as well as other commercial systems have always run on fossil fuels impacting our environment. The transportation platform alone is responsible for 71% of all US fossil fuel use.[3] In fact, the contributions of the direct transportations sector to GHG and CO2 are also high (56%) as reported by Environmental Protection Agency (EPA).[4] EPA further reports that since 1990 the GHG emissions have increased by 7%, varying year to year according to the price of fuel, GDP growth, and other factors (Figure 6.1 shows EPA's measures as of 2014). Some offsets are possible as sinks by forests and lands but not enough to combat future damage. The parking industry has its share of these gas emissions while lots are built and also while people are cruising for an empty space (meter, lot, or garage or even commuting) during congested traffic, in addition to the normal use of electricity for lighting. The question arises – could parking lots reduce the environmental hazard? Yes, in several ways:

- Supplementing gas fueled energy with renewable energy (wind and solar)

- Reducing frequent city commuting via newly defined SPD and STPD

---

2 *News & Observer* article and Internet data.
3 U.S. Department of Energy, 'Bureau of Transportation Statistics, 2010 – Energy Information Monthly Review', in *Transportation Statistics Annual Report.* Washington, DC, 2010, 143.
4 EPA Climate Change Report, 'Sources of Greenhouse Gas Emissions', April 2016.

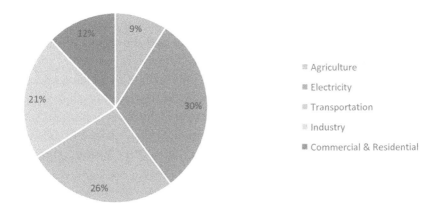

Agriculture
Electricity
Transportation
Industry
Commercial & Residential

FIGURE 6.1   Total GHG emissions in 2014 (6,870 million metric tons of CO2 equivalent).

- Charging electric vehicles while parked
- Optimizing energy use in all transportation sectors as well as associated facilities for self-sufficiency

Other usage of these lots while cars or trucks are sitting idle is to be devised – tire change, maintenance work, and detailing of vehicles including autonomous vehicles, for example. To be effective, proactive policy recommendations are required. An understanding of future transportation services where parking becomes the tail end is a must. Parking has not received attention for many years and technology, especially renewable energy, for the parking sector is ripe to take it to the next level. To make parking as useful partner to other transportation sectors, one needs to take advantage of its potential for producing energy in addition of other contributions. In this chapter, a model to measure the renewable energy potential will be described. Open spaces that may be oriented for sun or wind exposures will be candidates for such a model. One can then see the advantage in relocating all spaces later based on principles outlined in RWPTTP. This idea dawned in me when working on a project to establish a legacy of renewable energy in North Carolina as a member of the North Carolina Governor's board of science, technology and innovation (BSTI)[5] and an article in the *Triangle*

---

5 BSTI, (www.nccommerce.com/sti/board-of-science-technology-innovation).

Business Journal.[6] In addition, political and social pressures are pushing nationally and internationally to explore renewable sources of energy to preserve the sustainability of the environment and the quality of our life-styles, the 2015 U.N. Climate Accord, France (Paris).

For the stated goals, characterization of 900 million to 1 billion parking spaces for cars and 15 million trucks in the U.S., transporting goods to and from Mexico and Canada, using our interstate highways is important. Such characterizations are also under consideration by the Federal and different State Department of Transportation (DOT) 2015–2019 Vision.[7] One can characterize these parking spaces in many ways:[8]

- Over 30% to 40% of city's prime land is occupied.

- 10% to 15% of suburban land is occupied.

- Each car needs a space when they are not driven (95% of time).

- 160 billion square feet of concrete and asphalt have been used.

- Over all, the distribution is 30% on-street (metered or non-metered) and 60% off-street parking (surface or structured).

- 5% to 10% is under a structure mostly in the urban areas.

- There are not enough forests and trees in the open lot to absorb CO2.

- 10% of total transportation CO2 emissions are due to auto parking, the rest is due to highway driving, city driving, and traffic jams.

- Lack of a national parking database causes underutilization of space or no private and public cooperation.

- There is no uniformity for parking services nationwide for innovative new applications and services.

- There is no effort yet to declare parking spaces as national assets for future energy exploration.

---

6 'One Way to Narrow the Urban–Rural Divide', *Triangle Business Journal*, May 18, 2018.

7 ITS and Joint Program Office, 'ITS Strategic Plan -2015-2019'.

8 Mikhail Chester, Arpad Horvath, and Samer Madanat, 'Parking Infrastructure and the Environment', *Access*, vol. 39, Fall 2011.

In spite of the above environmental impact, there is no slowdown in constructing more parking spaces using the old formula because car sales are going up and up.[9] How can we escape this monstrous environmental destruction? It is high time for all of us to take actions or seek technological help. Yes, there is a trend to manufacture environmentally friendly electric cars, which do not use fossil fuels, but they cannot keep up with the demand in many developing countries such as China, India, and Brazil. It may take several decades to bridge this gap. In addition, these smart cars (self-driving, self-parking, and auto collision prevention), if all electric, will require energy for charging. The new phenomena point us towards a new parking paradigm to keep up with technological changes.[10] The proposed paradigm should be made an integral part of ITS. Here, I am proposing a smart way to reutilize the vast land of parking spaces as source of renewable energy[11] to add sustainability to the transportation infrastructure. The pioneering work has been initiated by Lockheed Martin in Florida. Lockheed Martin in collaboration with the Florida Governor and Advanced Green Technologies covered 151,400 square foot parking areas (500 spaces) with solar array in September 2015. The upshot is the creation of a 2.5 megawatt capacity solar carport to supply energy to their facility instead of grid-supplied energy using fossil fuel, coal, gas, nuclear, or even hydro. This solar car port has the potential to generate 3.33 million kilowatt hours per year renewable energy, saving millions of dollars over its lifetime. In addition, it will help Lockheed Martin's goal of reducing gas emissions by 35% according to their research data, if implemented in all their locations. Such consciousness of renewable energy will also help to reduce energy use, with many peripheral low energy sensing devices including Light Emitting Diodes (LED) and remote wireless control systems using smartphones and other mobile devices.

Open Space Estimate

Exact estimates of all existing open lot spaces are difficult to ascertain. A reasonable estimate of open lot spaces that are suitable for renewable energy is required. Extrapolations of field data available from different

9 D. Shoup, 'The Trouble with Minimum Parking Requirements', *Transportation Research A,* vol.33A, no.7/8 (1999), pp 549–74

10 Amalendu Chatterjee, 'Parking Part of ITS – Why and How?', *Parking Today,* March 2016.

11 'Lockheed Martin Parking Catches Sun Power', press release.

sources with assumptions made to compensate for the lack of historical record keeping nationwide. I could take the total number of 900 million spaces and then assume variations from as low as 10% to as high as 80% as open spaces to support my hypothesis. But I decided to be more scientific. I looked at meter spaces (that are easily relocatable) and open spaces as they exist (not garage spaces). They are considered first because they are the best candidates for renewable energy. Once this hypothesis is proven, one can extrapolate garage spaces (if these are required in the new parking paradigm) later once all parking spaces are relocated from city business center (CBC) to suburban parking district (SPD). Garage spaces to SPD will be planned as open space eventually for a greater benefit to the transportation industry.

METER SPACE ESTIMATES

Data from an IPI presentation[12] for twenty-one cities were extrapolated to get an estimate of total meter spaces nationwide, including some assumptions. The IPI presentation had real data of typical cities by meter space, population, revenue from meter, and revenue from tickets. Summary of the IPI counted data is as follows:

Total installed meters in 21 cities:	208×10^3
Population of 21 cities:	28×10^6
Numbers of incorporated cities in U.S.A.:	19,354
Total US population:	325×10^6
Average meter per city:	$208,000/21 = 9,904$
Average meter per population:	$208,000/28 \times 10^6 = 0.0074$

Each average above translates to $(9904 \times 19354) = 192 \times 10^6$ or $(325 \times 10^6 \times 0.0074) = 2.4 \times 10^6$ meter spaces respectively for all US cities. There is a big disparity but given the wide range of city size and population (rural vs. urban) I will assume the average of the two which is 97.2×10^6 as the total meter numbers. Theses data is within the range of similar research statistics floating in the industry.[13] Further analysis below will show the potential capacity of those spaces for renewable energy generation. Other benefits from these spaces if relocated to the suburban areas as part of RWPTTP

---

12 IPI Presentation by L. Dennis Burn of Kimley-Horn and Michael Klein of Klein Associates, May 2016.
13 Aashish Dalal, January 30, 2014, PANDO.

parking paradigm will be considered in other chapters. For example, the possible solar energy capacity from 96.7 million relocated meter spaces if covered by PV cells may be as high as (assuming each 500 spaces generate 2.5 MW occupying 3 acres for each MW):

$$97.2 \times 10^6 \times 2.5 \text{ MW}/500 = 0.486 \times 10^6 \text{MW} = 486 \text{ GW}.$$

If those spaces use wind power, the wind energy capacity from those 204 million spaces may be as high as (assuming occupancy of 30 acres for each MW).

$$97.2 \times 10^6 \times 2.5 \text{ MW}/5000 = 0.0486 \times 10^6 \text{MW} = 48.6 \text{ GW}.$$

Each wind turbine occupies 10 times more acres on average for generating each MW of energy. Solar capacity from meter spaces is almost close to half the amount of the grid capacity now (1074.3 GW).[14] We can take advantage of this if all meter spaces are relocated conveniently in suburban areas and autonomous transportation services are introduced to the urban city. Of course, we have to account the fact that some locations are suitable for solar only and some for wind only.

Open Lot Space Estimate (Existing)

These spaces are mostly located at the outskirt of cities. For the simplicity of a better estimate, they are divided into ten different facility types as shown below. There may be more facility types beyond those listed below and that also may be of significance. Each facility type will have different ratios of open vs garage spaces. I will describe our method to derive the estimate. Hopefully, a reasonable but conservative conclusion can be derived from these data.

1) airports

2) suburban and urban commercial and office buildings

3) shopping malls

4) sport complexes

5) grocery stores (national chain)

---

14 Energy Information Administration (eia), Wikipedia, 2018.

6) universities

7) hospitals

8) hardware stores (national chain)

9) concert arena

10) truck parking

Again, rather than using empirical formulae, a scientific method was used with some assumptions. The author visited each facility type in North Carolina and made physical counts of all spaces. The percentage (%) of these facilities are then applied to the national data. It is a limited statistical sample but very typical of similar cities. Big cities may differ substantially but on the average it may balance out given the large number of facility types that exist nationwide. Table 6.1(a), (b), (c), (d), (e), (f) gives raw data of facility type data and the percentage of open spaces, as collected by the author. Such characterizations and reorientation of these parking spaces (for truck or cars) are important to explore the full potential of renewable energy by photo voltaic (PV) cells or wind turbines. PV cells' production of energy and proposed orientation by SPD or STPD will bring economic growth. SPDs and STPDs may be strategically located to reduce the rural and urban economic gaps[15] while contributing to general growth. Wind power in certain geographic areas may also add certain flexibilities to the new way of thinking while reconstructing new parking facilities in coordination with other transportation services. All truck spaces are in open lots, hence easily adaptable for renewable energy generation.

1) Airports

In the U.S.A., there are over 19,700 private and public airports according to the 2011–2015 National Plan of Integrated Airport Systems (NPIAS). Of these, 5,170 airports are open to the general public and 503 of them serve commercial flights. I have compiled data from forty major airports. Parking spaces for these airports are estimated by the standard formula based on numbers of enplaned passengers (three spaces per 1,000 originating passengers).[16] The largest airport (Atlanta)

15 Amalendu Chatterjee, *Triangle Business Journal*, May 2018.
16 IPI, 'New Report Highlights Importance of Parking to Airport Operation', press release.

TABLE 6.1 RTP's Collected Field Data by Facility Type for National Total

(a) Airport				(b) University			
Facility	Total Parking Space	Open	% of Open	Facility	Total Parking Space	Open	% Open
RDU	19,277	7785	40	NC State	20,000	12,980	64.9
Charlotte	34,500	27,500	80	UNC	22,240	3,142	14.12
Piedmont	4,000	2,000	50	Duke	23,033	12,005	52.12
Total	57,777	37,285		Total	21,758	9,376	
Average	19,259	12,428	56.67	Average	7,253	3,126	43.1

(c) Sports Complex				(d) Shopping Center			
Facility	Total Parking Space	Open	% of Open	Facility	Total Parking Space	Open	% Open
PNC Arena	8,000	8,000	100	South Point	6,400	6,400	100
Kenan Stadium	1,500	300	20	Walmart	3,000	3,000	100
Carter Finley Stadium	7,000	7,000	100	Target	2,000	2,000	100
Total	16,500	15,300		Total	17,060	17,060	
Average	5,500	5,100	73	Average	5,687	5,687	100

(e) Hospital				(f) Concert Arena			
Facility	Total Parking Space	Open	% of Open	Facility	Total Parking Space	Open	% Open
Duke Wake Med	960	0	0	Koka Booth	6,400	6,400	100
Rex	2,458	658	26.76	Duke Energy Center	1,852	1,228	66
Cary Wake Med	800	350	43.75	DPAC	3,000	0	0
Total	4,218	1,008	23.5	Total	5,994	2,370	55

out of those forty has over 138,000 spaces and the smallest one has close to 15,000 spaces. These spaces are either in garages or remote open lots. It is assumed that these open spaces are always in a cluster containing more than 300 to 500 spaces per lot for transportation efficiency of passengers to the terminal. Transportation from those open lots is provided free by the airport authority. The author has counted parking spaces of three airports (Charlotte, Raleigh, and Piedmont) in North Carolina. Average spaces for those three airports are 19,000 (garages and open lots) and the weighted average for open lots is 57%. A cluster of 500 open spaces can generate 2.5 MW (Lockheed Martin's model of the Florida office building).

Number of airports open to the public considered = 5170
Total number of spaces in U.S.A. airports = $5170 \times 19 \times 10^3 = 98 \times 10^6$
Number of open spaces qualifying for PV cells = $98 \times 10^6 \times 0.57 = 56 \times 10^6$
Total PV energy generating capacity = $56 \times 10^6 \times 2.5/500 = 280$ GW
Total wind energy generating capacity (1/10th) = 28 GW

Many airports can convert all possible parking spaces (shown in Figures 6.2) to PV or wind turbine generating stations depending on the location. Such efforts could benefit airports as well as their customers in different ways. All electric cars will need plug-ins and one way to do it cheaply is to use renewable energy for the convenience of passengers who will be out for an average of three to four days. Such conversion plan would add values in three other main areas in addition of saving electricity costs: reduction of fossil fuel use (GHG emission), supplement of airports' general energy requirement during the peak hour demands, and off-peak hour loads can be sold to the utility company. Utility companies have three different rate structures: Residential, commercial, and industrial. Examples of parking lots chosen here will fall under commercial rate. Again, commercial rates in some regions are higher than others justifying faster rate of return on new technologies.

2) Suburban and Urban Commercial and Office Buildings
There are too many variables to correctly assess the number of parking spaces under this category. Number of tenants sharing the building, number of employees by each tenant, type of business, locations (urban or non-urban), and the latest trend of square foot rented per employee (traditional 4 per 1,000 square feet to current 2.5) determine number of parking spaces. On the top, parking spaces are shared by all tenants with no fair allocations between them.[17] In urban cities, all parking spaces are mostly in garages in contrast to suburban areas. According to one report, there are $84,096.3 \times 10^6$ square feet of office spaces in the U.S.A.[18] Another report reconfirms a parking rate per thousand square

---

17 Cresa Atlanta Blog.
18 Andrew C. Florence, *Slicing, Dicing and Scoping the Size of the US Commercial Real Estate Market*, CO-star Group, April 26, 2010.

FIGURE 6.2(a)   Typical airport parking lot.

FIGURE 6.2(b)   Parking configuration of office complex.

FIGURE 6.2(c)   Parking configuration of street parking.

FIGURE 6.2(d)   Parking configuration of a shopping center.

feet of office space.[19] Following the Gross Facility Area (GFA) guide-lines as per parking requirements[20] (2 spaces per 1,000 square feet of office space), one can estimate total parking spaces to be close to $84,096.3 \times 10^6 \times 2/1000 = 168.2 \times 10^6$. The author did not find a suitable complex to verify the open vs. closed spaces ratio in RTP, NC. It is safe to assume 30% spaces are in the open lot (conservative estimate) if we consider complexes are distributed equally between urban and suburban areas.

Total estimated parking spaces for suburban and urban commercial and office buildings = 168.2×10^6

Open spaces suitable for renewable energy (assuming 30% of the total) = $168.2 \times 10^6 \times 0.3; = 50.46 \times 10^6$

19 Brian V. Church, 'Trip Generation and Parking Demand for Small Office Complexes in Montana', MSU-ITE Student Chapter President Report, May 28, 2009.
20 Wikipedia.

Total PV energy generating capacity = $50.46 \times 10^6 \times 2.5/500 = 252.3$ GW
Total wind energy capacity = $252.3/10 = 25.2$ GW

3) Shopping Malls

Parking in a shopping plaza can be an experience during holidays (Christmas and Thanksgiving). You circle many times before you can find one shopper leaving and you line up before anybody else grabs it. The rest of the time it remains open (rain or shine) and empty. Numbers of total parking spaces surrounding the mall depend on the total square foot of the covered space for retail stores, with some designated handicapped spaces. In big cities, shopping malls have covered garages but in the suburban malls, parking spaces are open. There are over 116,000 malls in the U.S.A., each having 10 to 200 stores with some anchor tenants.[21] There has been a steady decline of shopping malls in the U.S.A. from 1996 till today @ 3% due to the increased online shopping. Some have covered parking spaces but even then the top floor is open where PVs can be installed. Data have been compiled for thirty-four large and medium shopping malls in different states. In addition, the author has physically counted available parking spaces of three average sized shopping malls in North Carolina, which are all in open spaces. The average of these three malls is 4,834 and they are all in clusters qualifying for solar PVs or wind turbine.

Average parking space per mall = 4,834
Total open spaces (nationally) = $4,834 \times 116 \times 10^3 = 561 \times 10^6$
Total PV energy generating capacity = $561 \times 10^6 \times 2.5/500 = 2805$ GW
Total wind energy capacity = $2805/10 = 280.5$ GW

4) Sport Complexes

In the U.S.A., there are over 240 stadiums with sitting capacity ranging from 18,000 to 107,000. Some of them follow standard guidelines of seat to parking space. But some make different transportation arrangements on the game day, such as having free park and ride for the ticket holder from neighboring areas where parking is available. The author has counted parking spaces of three prominent stadiums in RTP, NC: PNC arena, Kenan Stadium, and Finley Carter Stadium. From these counts,

---

21 Statista: Statistics Portal, 2017.

the average number of open spaces per stadium was found to be 15,300, out of the total 16,500 spaces. This provides a 92.7% percentage for the national counts. The potential of renewable energy capacity is as follows from these spaces:

National total spaces of sports complex = $240 \times 16,500 = 3.96 \times 10^6$
National open spaces of sports complex = $240 \times 16,500 \times 0.927 = 3.72 \times 10^6$
Total PV energy generating capacity = $3.72 \times 10^6 \times 2.5/500 = 16.6$ GW
Total wind energy capacity = $16.6/10 = 1.66$ GW

5) Grocery Stores (National Chain)
Several store types have been surveyed personally by the author in North Carolin :Target, Walmart, Whole Food, and Costco. Target had 490 open spaces, Walmart had 690, Whole Food had 220, and Costco had 420.[22] There were additional spaces to different strip malls along with those shopping centers but they were excluded while counting even though they may also qualify for renewable energy.
Total number of Target stores in the nation = 1,802
Corresponding open parking spaces in Target = $1802 \times 490 = 883 \times 10^3$ (i)
Total number of Walmart stores in the nation = 4,177
Corresponding open parking spaces in Walmart
$$= 4,177 \times 690 = 2,882 \times 10^3 \qquad \text{(ii)}$$
Total Number of Whole Food stores in U.S.A. = 470
Corresponding open parking spaces in Whole Food
$$= 470 \times 224 = 105 \times 10^3 \qquad \text{(iii)}$$
Total number of Costco stores in the nation = 741
Corresponding open parking spaces in Costco
$$= 741 \times 420 = 311 \times 10^3 \qquad \text{(iv)}$$

Total Adding (i), (ii), (iii) and (iv) = $4,181 \times 10^3$
Total PV energy generating capacity of the above
$$= 4,181 \times 10^3 \times 2.5/500 = 20.91 \text{ GW}$$
Total wind energy capacity of same = $20.91/10 = 2.1$ GW

---

22 Wikipedia.

6) University

Three universities in RTP were surveyed, counting the total parking and open spaces. The average of parking spaces in these three universities is 21,758. The weighted average is 43.1% for open spaces, an average of 9,376. These three universities (Duke, NC State, and UNC-CH) are typical of the national average. Out of these three, UNC-CH claims its parking spaces short fall of the total requirement but they are supplemented with free park and ride transportation services from neighboring facilities.

Total universities in U.S.A. = 4,583
Total campus parking spaces nationwide = 4,583 × 21,758 = 99 × 10^6
Total campus open spaces = 4,583 × 9,376 = 42 × 10^6
Total PV energy generating capacity of the above
$$= 42 \times 10^6 \times 2.5/500 = 210 \text{ GW}$$
Total wind energy capacity of the above = 210/10 = 21 GW

7) Hospitals

In the U.S.A., there are 7,600 hospitals. The top 100 of them have 86,100[23] beds in total, requiring close to three parking spaces per bed. The author has counted total spaces (open spaces and garages) for three hospitals in RTP, NC. They are Wake Med in Cary, Rex in Raleigh and Duke Med in Raleigh, consisting of 208 beds, 665 beds and 171 beds respectively. They are typical in meeting the national average of bed to parking ratio. Counted total numbers for each are 800, 2,458, and 960 respectively. The average spaces per hospital are 1,406 and the weighted average to open to closed space is 24%. Applying this average to the national data we can calculate:

Total number of hospital parking spaces (nationwide)
$$= 1406 \times 7600 = 10.7 \times 10^6$$
Total number of spaces suitable for PV cells or wind turbine
$$= 10.7 \times 10^6 \times 0.24 = 2.6 \times 10^6$$
Total PV energy generating capacity of the above
$$= 2.6 \times 10^6 \times 2.5/500 = 13 \text{ GW}$$
Total wind energy capacity of same   = 13/10 = 1.3 GW

_____

23 www.beckershospitalreview.com

8) Hardware Stores (National Chain)

Home Depot and Lowe's have been surveyed by the author, counting the exact number of open parking spaces in Research Triangle Park (RTP), North Carolina. Data for each store were as follows:

Parking spaces at one Home Depot = 656 (a typical sample)
Parking spaces at Lowe's   = 480 (another typical sample)

In the U.S.A., there are 2,000 home depo stores and 2,394 Lowe's stores. Extrapolating from the data we can estimate the total number of open parking spaces as follows:

Total open parking spaces in Home Depot (nationwide)
$$= 656 \times 2000 = 1,312 \times 10^3$$
Total open parking spaces in Lowe's (nationwide)
$$= 480 \times 2394 = 1,150 \times 10^3$$

Total open spaces for the hardware stores (nationwide) $= 2,462 \times 10^3$
Total PV energy generating capacity of the above
$$= 2.5 \times 10^6 \times 2.5/500 = 12.5 \text{ GW}$$
Total wind energy capacity of same = 12.5/10 = 1.25 GW

9) Concert Arena

There are over 328 concert arenas in the U.S.A. with seating capacity ranging from 750 to 23,500. Some are open and some are indoor, like the most stadiums. Parking spaces are also divided into garages and open lots. The author visited three of them in RTP, NC, and counted parking spaces designated for those arenas. The average of the three counted is 1998 per concert hall and the percentage of open space is 55%. Applying this to the data for the national total:

Total number of open parking spaces for the concert halls
$$= 328 \times 1998 \times 0.55 = 360.4 \times 10^3$$
Total PV energy generating capacity of the above
$$= 360.4 \times 10^3 \times 2.5/500 = 1.8 \text{ GW}$$
Total wind energy capacity of same = 1.8/10 = 0.18 GW

10) Truck Parking

Each truck parking space occupies three times more square feet than the passenger car on average. There are close to 5,000 truck stops in the U.S.A. This may translate all existing rest areas and private truck spots close to 1 million equivalent car parking spaces for the source of potential renewable energy. For 15 million trucks, the future space requirements will be much higher than 15 million spread over in all interstate highway corridors, assuming each truck will rest for certain times of the day. The ratio may not be as high (3:1 or 5:1) as private vehicles. For the exercise of this chapter a 1:1 ratio (truck to parking space) has been assumed.

Total estimated parking space (interstate highways) = 15×10^6
Total PV energy generating capacity = $15 \times 10^6 \times 2.5 \times 3/500 = 225$ GW
Total wind energy capacity = $225/10 = 22.5$ GW

We can now summarize the above finding in Table 6.2 to validate assumptions and data nationwide. Assumptions made to calculate total number of parking spaces are reasonable, though the number came out higher than the floating estimate in different market reports. That raises the possibility of fine tunings of data collections in the field. The goal is to assess the extent of tremendous hidden resources for the sake of the environment and different reuses of land as appropriate. Even discounting these numbers substantially the potential of renewable energy is almost four times the capacity of the current grid (4,446 GW vs. 1073 GW) mostly energized by fossil fuels. Out of all facility types, shopping mall spaces alone stand out as producing almost double the current capacity of the grid. This capacity, along with the reduction of total energy consumption by using smart lighting systems, may mitigate national and international climate concern. That analysis is beyond the scope of this book.

How do we exploit such hidden gems for the national interest? Decision makers and policy initiators must act quickly and properly. For example, California State legislators have made it mandatory for all new homes built after 2020 to be self-equipped with solar PV cells. I can point out four areas of policy regulation in this context but there may be more.

(1) No more construction of new parking spaces until existing ones are fully utilized,

(2) Relocation of meter spaces and garages from the city center to the suburban areas,

TABLE 6.2 Possible Energy Capacity by Parking Facility Type

Parking Facility Type	National Total Number of Facilities	Total Parking Space	Open Lot Percentage Based on RTP Facility	Total Open Lot Spaces	Possible Energy Capacity By PV Cells	Possible Energy Capacity by Wind Turbine
Relocated Meters	19.35×10^3	97.2×10^6	100%	97.2×10^6	486 GW	48.6 GW
Airports	5.17×10^3	98×10^6	57%	55.86×10^6	280 GW	28 GW
S&U Office Complex	824×10^3	168.2×10^6	30%	50.46×10^6	252.3 GW	25.2 GW
Shopping Malls	116×10^3	561×10^6	100%	561×10^6	2805 GW	280.5 GW
Sports Complexes	240	3.96×10^6	92.7%	3.67×10^6	16.6 GW	1.66 GW
Grocery Stores	6.45×10^3	4.18×10^6	100%	4.18×10^6	20.91 GW	2.09 GW
Universities	4.58×10^3	99×10^6	43.1%	42.67×10^6	210 GW	21 GW
Hospitals	7.6×10^3	10.7×10^6	23.5%	2.5×10^6	136 GW	13.6 GW
Hardware Stores	4.4×10^3	2.462×10^6	100%	2.462×10^6	12.5 GW	1.25 GW
Concert Arena	328	0.655×10^6	55%	0.36×10^6	1.8 GW	0.18 GW
Truck Parking	5.0×10^3	15×10^6	100%	15×10^6	225 GW	22.5 GW
Total	975.12×10^3	$1,167.15\times10^6$	80.7 %	942.16×10^6	4,446 GW	444.6 GW

(3) Private and public cooperation for consolidated parking spaces, and

(4) Introduction of solar PV cells or wind turbines in all convenient parking locations.

Once carefully planned, such exploitation would also reduce gas emissions due to fossil fuels and enhance other uses of free electricity such as charging all electric cars while parked. Possible improvements of three parameters are graphically shown in Figure 6.3: power generation, energy cost saving at two different rates, and GHG reduction. Energy saving costs will depend on the available electricity rate. Reduced energy usage will depend on the sophistication of control systems and sensing devices installed in the total power infrastructure. According to the 2016 limelight conference show by San Diego Conference Center and TwistHDM Corporation (www.twisthdm.com) at least 50% energy savings

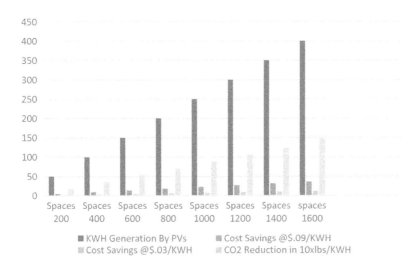

FIGURE 6.3   Energy cost savings and GHG reduction by PVs.

are possible if the control system can leverage occupancy sensing, daylight harvesting, event scheduling, and proper zoning to use energy only when, where, and to what degree it is needed.

CONCLUSION

The above research presents a cursory look at the potential. The findings may reinforce the resurgence of American manufacturing in renewable energy, proving the statement by Ralph Waldo Emerson, the famous poet and essayist of the 19th century, 'America is another name for opportunity'. Imagine over 900 million parking spaces being used to generate electricity, with a capacity of 4.5×10^6 MW or 4500 GW – more than the current grid capacity. Yes, 100% may not be ready right away but there is the potential for them to be a bigger source of energy than any other alternative method. Even if one-third of 900 million parking spaces (open lots) are used, it will generate over 1.5×10^6 MW = 1500 GWs energy, almost more than today's US power grid infrastructure capacity.[24] It will be a serious social crime if such an opportunity is let pass without proper consideration. Of course further refinements of the data presented will be required before full implementation is possible

---

24 Wikipedia.

when government regulation is favorable. Such efforts come with other associated benefits which we could not have dreamed of before – cost reductions, saving nature, promoting electric vehicles, and protecting vehicles from extreme heat and cold. There is some concern when the sun is not shining. Wind power is an alternative. There are also back up batteries available to take care of such adverse situations. It will increase the cost. There are other backup energy storage technologies such as thermal energy, compressed air, and Ice Bear Storage competing with the battery energy storage backup. The good news is that the cost of all cutting edge battery technologies is coming down with higher efficiency. The battery technology is also changing from lithium-ion to zin-air[25] to be lighter and safer. It is time to plan for the future by unlocking the hidden power of parking assets.

The International Parking and Mobility Institute (IPMI) has instituted an ITS-Parking task force to link parking services with ITS services[26] and also Alliance Parking Data Standard (APDS). The author is a member of this task force. Renewable energy sources may be another element fot it to consider while exploring how to raise the awareness of the two vested communities. The solar/wind power source and its backup have the potential to act as isolated portable grid to improve security from hackers during a national emergency.

25 'Cheaper Battery Is Unveiled as a Step to a Carbon-Free Grid', *NYT*, September 26, 2018.
26 Jason M. Jones, 'Boosting Intelligence: An ITS Task Force Takes on Intelligent Transportation Systems and How they Might Work with Parking, and Members Want Input', *Parking Professional Magazine*, February 2016.

CHAPTER 7

Enhancing IPMI Sustainability Framework

INTRODUCTION

The International Parking and Mobility Institute (IPMI) has always been active in bringing end users, vendors, and other stakeholders together. With that in mind, IPMI started some serious thinking on a Framework on Sustainability for providing end user services. The framework suggests some guidelines for Parking Design, Management, and Operations. There are ten stated Goals related to the Framework and an Action Plan for putting the Framework into action was published in January 2012. The plan is a bold step forward for the parking industry. IPMI's foresightedness will make a difference and its commitment to continuous evolution of this framework is equally praiseworthy. We owe a lot to the efforts made by these people for the future of the industry. This creates, of course, an opportunity for all of us to contribute. In this chapter, attempts are made to bring new thoughts to enhance the framework. Additional action items and the reasonable means to pursue an achievable plan are identified. To achieve a reasonable result, one must take into account many internal and external factors in the IPMI's plan:

- Web Portal for Parking Services (mobility and flexibility with software such as iPhone APPs) – one application for all parking scenarios.

- Parking Services being part of Intelligent Transportation Services (ITS) and Public, Private Partnership (PPP).

- Hardware independent regulations to ease technology exploration, such as promoting a Web Portal to reduce cash and audit burdens related especially to Parking Access and Revenue Control System (PARCS).

The author believes outdated regulatory issues at all levels of the industry are bottlenecks for technological breakthrough and VIP like customer services. Also, industry thinkers and willing participants must take some risks for implementation and deployment of what they preach. Hopefully, these thoughts will enhance the work done so far by IPMI.

BACKGROUND AND REFERENCES

In addition to IPMI's perspective (the insider view), there are many external views, expectations, and events that need a serious investigation and analysis to be factored into the Framework on Sustainability. This is the theme of the chapter. For example, it is very difficult to find a day in the news media that goes by without some incident on parking (good or bad), as already discussed in Chapter 1. Conclusions that can be drawn from these can be summarized into few bullets:

- The single-space parking meter does not have a long-term future – it is clumsy and does not adapt to newer technologies.

- Shaming a customer with a Sticker of Shame (that cannot be removed easily) for parking violation is not a norm for business courtesy.

- There are at least one billion parking spaces in U.S.A. One-third of them are in the parking lot making four parking spots per car and three spots per person. In some US cities, parking lots cover more than a third of the land area. Of course, urban and suburban lots are not the same. Outmoded zoning codes requiring specific ratios of parking spaces from the auto-boom days should be abandoned, in favour of more efficient public transportation and sharing of scarce resources as part of a common pool.

- The National Association of Counties projects that between $10 and $40 billion in county fees and fines go delinquent every year in U.S.A. That figure does not include municipal citations or the cumulative amount of outstanding delinquent fees and citation from previous years. Why let it go uncollected if you can collect part or whole of it in advance? Efforts should be made to use modern technology to smooth the regulatory regime.

- Google driverless car uses artificial intelligence software to drive or to park. Legislation needs to be prepared so that driverless car technology can resolve human error including falling asleep at the wheel or driving drunk.

- Tyler Cowen of George Mason University says that our economic downturn has been, at least in part, the result of slow technological progress, combined with a century ago. The next age of car computers will enable cars that are aware of their own location and the location of other vehicles to 'self-organize'. They will talk to one another and the new deployed infrastructure in order to optimize traffic flow, minimize congestion, reduce pollution, and increase general mobility. Imagine a future in which even a 90-year-old person can remain mobile over long distances in a car that drives or parks itself. That may mean more visits from grandparents – and it might also mean that the car could drive them straight to a hospital in a medical emergency. According to Conor Friedersdorf of Atlantic Tech, driverless cars could mean the end of parking – instead, we could collectively own far fewer cars that are always in use through a sharing approach.

- Customers should be getting VIP treatment from a service industry like the parking industry. Cities providing such services should feel proud not intimidated. Days of meter specific or dependent parking are numbered. Instead, customers should have a day's pass – free to park anywhere he or she prefers using mobile phone and PIN code. Customer must know where they are parking before they start the car so that idling and circling in garages and airports are reduced for the sake of controlling CO_2 emission and traffic congestion.

- Technologies are available to change the paradigm of parking fee collections using a Web Portal. In fact, the paradigm could be reversed from delinquent collection to advance collection. The Web Portal can also introduce civility into the parking industry like any other city services where you are not declared a criminal for a one-minute service violation. The criminal courts are flooded with traffic violation cases consuming unimaginable resources. Some declare them as ill-gotten money and there are many ways to claim them back. But the bottom line is: why not pursue other means to get the same end result from the beginning, not by collection agencies or traffic courts? Many cities are devising incentives to collect unpaid tickets at discounted rates.

- In most cases especially in NYC, lost public parking spaces are not replaced, because zoning rules discourage developers from adding parking to new residential buildings. Will that discourage car ownership? Answers are in the details of how parking and intelligent transportation services (ITS) are coordinated and played out in the hand of regulators – getting few tickets vs. paying what you would in a lot as rent. And also the Clean Air Act's stricter codes will limit the number of new parking lots in the city, reversing the percentage of off-street parking usage by commuters and residents.

- The latest article in *New York Times* is on how parking has been used as instructional guidelines for the kindergarten children.[1] Second graders to fourth graders went around the block to study Muni-Meters and parking signs. They learned new vocabulary words, like 'parking', 'violations', and 'bureau'. They also calculated how much they needed to put in for 40 minutes if the rate is 0.50 cents per hour. Imagine the potential when high technology including parking apps in iPhone is introduced for Generation Y. Such new apps could be training materials for high school or college graduates.

The changes overall are big enough to undermine many of the legal systems we rely on – from insurance to criminal traffic/parking laws. The question now arises for all of us – what we need to do and how fast we need to do those changes in the parking industry. Days of asking 'Why Change?' are gone. Changes are required to make parking better, elegant, and more acceptable to the modern society – a new way to do business for tomorrow. Let us talk about of some difficulties we face for

---

1 'A Field Trip to a Strange New Place: Second Grade Visits the Parking Garage', February 12, 2012.

making these changes. We must ask ourselves in spite of all such efforts why we have not made enough progress.

INGRAINED PROCESS, PRACTICE, AND PROCEDURES

The parking industry started in the early 20th century with a basic concept of making money by renting city properties on a time share basis. The simple concept of making money became more sophisticated with meter technologies and Parking Access Control and Revenue System (PARCS). City and County regulators took advantage of this opportunity to make more money by introducing the concept of violations, ticketing, and enforcements. Even today, the city budget is correlated with realized and unrealized parking ticket revenue collection. For example, the Office of the Comptroller, New York City, publishes actual and budgeted parking fine revenues in its Comprehensive Annual Financial Report (CAFR). Actual revenue reported for the past three fiscal years is over several hundred million dollars and includes parking fines and red light camera monitoring violations. The CAFR provides several budget categories. *Adopted Budget* refers to estimated parking revenues projected prior to the start of a fiscal year as part of the New York City budget approved by the City Council. *Modified* refers to revised projection made before the close of the fiscal year as part of quarterly modifications to the City. All these ingrained processes and procedures (regulations) may be difficult to undo for the sake of modernization. Unfortunately, these processes and procedures are no longer up to the mark when technologies can simplify fee collections on regular hours, over time hours, special events, and other customized home-printed or cellphone reserved and permit options. Many of the existing factors are manmade, labor-intensive and could be attributed to human psychology or habit. A good reasoning and logical explanation can go a long way to solve many of those myths. Change is always difficult but must be accepted for a brighter outcome.

There are many myths surrounding the change. One must look at them critically to succeed in defining a new paradigm with its benefits and advantages.

- People's mindset for a change
- Easing regulatory dilemma

- Cost improvement (not capital alone but yearly maintenance)
- Lack of national uniformity and standard for parking

People's Mindset for a Change

There are many opposing forces playing their roles. I assume public facilities are victim of those forces. For example, parking is always an after the fact issue to local politicians. Old ladies with no cellphone in their hands may be happy to put coins in and park for one or two hours for a cup of coffee. Generation Y, geared to the latest electronic gadgets, has no patience with old technology or coins and on the spot payment. Generation Y will be happy with mobility and flexibility of payments and pay-as-you-use service options. How do you come up with a common solution that will also make the old lady happy? Instead of a common solution, city leaders are installing sophisticated Pay & Display meter technology (new version of the old concept) increasing the inventories, cost of operation, and maintenance. It also adds another piece of hardware to the system (credit card) and networking complexity.

Here is a challenge for all of us. For the old lady things got complicated even with the meter. Now, she has to walk a few feet back and forth from the meter to the car to pay and display the receipt irrespective of rain or shine. Should we favor Pay & Display machines over meterless parking with soft meter application (SMA™)? With the soft meter, the old lady can use her cellphone sitting in the car, identify herself to the system, and park in a marked space. She does not have to walk several feet to activate her parking privilege. It will be easier to train the lady to use cellphone than to understand all the buttons and logic of the Pay & Display machine.

Easing Regulatory Dilemma

Regulatory ease and adjustments are required in all seven areas of the Framework on Sustainability. One can have a better interpretation of each of those areas when regulatory issues become the founding pillar.

a) *Public/Private Partnership (PPP)* – All resources should be placed in a pool for resource sharing and higher utilization. It will be a win-win situation for all parties. It will also help cost effective use of technologies and give a good image of a city with a user- and business-friendly environment.

b) *Ticketing* – The idea of extra revenues from ticketing must be normalized. In other words, the parking industry must be decriminalized. If tickets are not paid, use of collection agencies is not a civilized business model except in extreme cases. There are many other means to collect 'over time' or 'high rate' payments from violators. No other service industry does this way. Why not change it? Even NCDOT has introduced an honor system toll collection with the help of high-speed cameras in addition to RFID high-speed tags. It could be replaced by a web portal with an account for every user, like any other city service, where fees could be paid in advance. Instead of ticketing when the time limit expires higher rates could be charged to the account or the credit card in accordance with other service business models – no more fist fights or use of criminal justice resources for ticket collections.

c) *No Meter in the Street* – Do you still need the old meter technology (single or sophisticated Pay & Display) on the street? It does not mean you have to replace them all at once. Why not ease regulations to make room for the soft meter application (SMA™) or yet a better solution (RWPTTP) using smartphone and space identification or GPS auto identification to work with the hard meter technology or meter number? This is a transition plan leading to an eventual future system where an 800 number call (iPhone) can work as a meter. The benefits of such strategies are that there is no lead time to generate new revenues, no digging up the street, no inventory, no maintenance, and no cash collection.

d) *One Unique Web Portal* – This would be a hosted system for all parking scenarios such as street parking, garage parking, and lot parking where work responsibilities between the administration and the driver could be shared like in the airline system (home-printed pass and bag check-in). This will fold many diverse systems into one. There will be one system for data preservation, the collection process, permit issuance, and space reservation. Benefits are: knowing where to park before you start the car, making changes as your schedule changes, one system to maintain as part of IT infrastructure, and above all a 'Go Green' implementation to reduce CO2 emission and traffic congestion. According to IBM research, findings in the area are as follows:

- In LA Business District (five square miles), driving around in search for parking in one year generated the equivalent of thirty-eight trips around the world. It burns 47,000 gallons of gas and emits 730 tons of CO2. It takes over 10,000 years for CO2 to come out of the atmosphere.

- Congested roadways cost $78 billion annually, with $4.2 billion lost in productive man-hours and $2.9 billion of wasted gas.

- 44% of CO2 emissions are due to driving and a percentage can be attributed to driving in the parking garages and lots in U.S.A.

We are facing a nightmare due to non-management and non-coordinated activities between the parking and the transportation industries. Parking services need to be integrated with the Intelligent Transportation Services (ITS) platform. The Framework on Sustainability should accommodate some of the ITS platform elements and address many regulatory issues in this context.

Cost Improvement

We are in an economic downturn. In spite of this, cities are spending millions to replace meters which may be obsolete in a few years. They do not have time or resources or political freedom to explore what is available to save money. There are two areas of cost savings required – capital investment to buy hardware and yearly maintenance of hardware. For example, meters are expensive and need many months of lead time. Parking Access and Revenue Control Systems (PARCS) are in a similar category. Both are labor intensive. A person might be hired at $15 per hour to collect $10 from an installed meter. They operate on cash transactions requiring elaborate audit practice to manage slippage. Vandalism of equipment is very common also. Moving towards a web portal (cashless and meterless parking and web automation, report generation, online appeal practice) can save costs with online or advanced cash collections and no or little field maintenance.

Lack of National Uniformity and Standard for Parking

In the ITS definition, parking resources are to be considered as national resources. Best utilization of these resources is through networking like the power grid, telephone and Internet. This sort of networking will motivate for public and private partnership and new innovations of revenue increase for all stakeholders. When I am traveling I would like to make one-click reservations for airline, hotel, car, and parking nationally and internationally. Parking companies may need to partner with Priceline, for example, to explore the business opportunity. In addition, there come parking apps Stores with iPhone – one click for satisfying your parking needs ahead of time or on the fly.

ENHANCING THE FRAMEWORK ON SUSTAINABILITY AND ASSOCIATED PLAN

The parking industry needs to be elevated to the next level in the civilized world of VIP services (with responsible work shared between the driver and the parking authority) with no cash in the field, no meter dependent parking, no inventory, no ticketing, no vandalism, and no traditional access control. IPMI's efforts are praiseworthy but may fall short in areas such as defining parking as part of ITS infrastructure; requiring new parking regulations influenced by new technology; identifying a proof of concept model (Web Portal or Soft Meter Application), and proactive implementation and deployment plan. This section presents items to add to all areas of IPMI's efforts – the Framework on Sustainability, Action Items, and Action Plan.

Framework on Sustainability

The Framework on Sustainability must have a foundation or a base where all seven items can further be defined, explained, modified, and expanded as and when needed. I believe the newly defined regulation base of the Framework can add much needed values and direction. 'Ease Regulatory Burden across the Board' (as shown in the Figure 7.1) is the theme of this regulatory base. Once the base is agreed upon, many thought-provoking ideas will prompt improvement. Parking regulations can be looked at as excessive in certain areas and lacking in others, especially with latest technologies. It also varies county to county and city to city. There may be central guidelines so that standards and uniformity can be achieved. Regulatory ease or new mode of operation is required to bring about technology deployment, meterless and cashless parking with an 800 call, account debiting instead of towing, no ticketing instead of increased fees, and uniformity and a standard way of parking whether you are in Raleigh, North Carolina, or in Los Angeles in California. Regulatory ease is also required for long-term goals such as Google's driverless parking, electric vehicles, stronger public and private cooperation for optimum resource sharing of mutual benefits, improving construction codes to maintain commercial/residential ratio to parking areas and 'Go Green' evolution strategy with solid CO/CO2 emission codes in cities including enclosed garages/lots. Specific questions for a new regulatory paradigm are as follows:

- Why issue a parking ticket? Replace it with an online account so that violators can be charged as in other similar city or utility services.

- Why have meter-specific parking? Try out meter independent but hours (or day) dependent parking with a home-printed parking pass.

- Why have handicapped decals or other decals? All car owners and their license plate information should be in the database for auto check for special consideration.

- Why take every parking offense to court? Introduce civility with the help of technology and clear documentation.

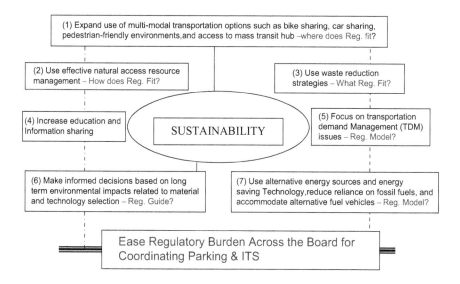

FIGURE 7.1   IPMI's modified framework with a base for sustainability.

- Why cash feed a meter for parking? Introduce an 800 call for parking activation/deactivation while sitting in the car so that cash collection can be avoided.

- Why have expensive meter, inventory, and field maintenance? Deploy a parking portal to replace meters with marked spaces and online collection.

- Why have idling or circling in the city lot or the garage? Make space reservations and avoid CO_2 emission and traffic congestion.

- Why not divide work responsibilities between the parker and the administration? Follow the model of the airline where you can print out boarding passes and luggage check from home or even just use the cellphone.

- Why not increase the mobility and convenience of parkers by extensive use of cellphones (iPhone in particular)? With simple registration and mobile PIN code, the parker has the flexibility to activate or deactivate parking privileges (street parking, garage parking, event parking) from any place, and at any time.

- Why do you need two separate processing systems (street parking and garage parking)? The web portal should be designed to integrate both.

- Why do you need the complicated Parking Access and Revenue Control System (PARCS)? Simplify it with less labor-intensive web portal applications (with easy audit) so that cellphone and the Internet are integral.

- Why not fold parking services under ITS? Joint planning will save resources.

- Why not connect all parking resources nationwide like the telephone network and electrical grid? Display the availability of parking space at the entrance to the garage.

- Why have office visits? Reduce the number of employees needed with online operation, account management, payment processing, auto payment processing, complaint/appeal handling practice, and other queries.

In essence, regulations will be the key that holds all pieces together.

Action Items

IPMI is a voluntary organization empowered by its members' input and strategy. This Sustainability Committee has expressed their vision in close collaboration with the members. Ten action items have so far been identified and more may be required. I have added three additional items (11 to 13) based on regulatory changes to the framework.

1. Developing seminars, webinars, e-learning opportunities, and speakers on topics related to sustainability, including a sustainability track at the annual IPI Conference and Expo.

2. Writing articles, white papers, and publications on the latest solutions that reflect a balance between economics, public health and welfare, and reduced environmental impacts.

3. Encouraging and recognizing achievements and improvements in sustainable parking and transportation through awards programs and exhibitor visibility.

4. Adding courses related to sustainability as part of the required curriculum for the Certified Administrator of Public Parking (CAPP) certification program.

5. Creating forums for peer-to-peer sharing on sustainable parking design, management, and operations

6. Developing tools and resources to facilitate sustainable design, management, and operations that result in long-term energy efficiency, informed material and technology selection, the availability of multi-modal transportation options, effective natural resource management, and the use of waste reduction strategies.

7. Forging partnerships with government agencies, vendors, and non-profit organizations to facilitate goal setting, information sharing, and funding incentives that encourage investing in sustainable parking solutions.

8. Supporting third-party certification solutions that recognize achievement in sustainable parking.

9. Initiating research that informs, justifies, and/or provides lessons learned on sustainable parking solutions.

10. Communicating with media, influencers, and the public to create awareness of the positive impact parking professionals can have on sustainability.

11. Identifying areas where viable regulatory changes and adjustments are possible for RWPTTP and ITS implementation.

12. Negotiating with the leaders of cities and airports to deploy a proof of concept model with new parking practice for a technology breakthrough.

13. Developing course materials for high school children and community college graduates on high tech parking such as apps in iPhone

Putting the Framework into Action:

Leaders must be aggressive in coming forward and implementing what they preach. They must also demonstrate a working model that brings benefits to the industry. Statements have been made covering many areas of technology. But they must work in sync to get the best result. One needs a revolutionary working and deployable model to move forward. I suggest the following:

- Try out the web portal on a trial basis in parallel with meters and PARCS to explore gradual but eventual replacement of complicated hardware.

- Select several participants who are willing to dedicate resources and funding for a sustainable working model.

- Accommodate as many features and function as possible available from the list above. Simplify and define a model that can satisfy many criteria requested by willing participants.

- Explore technology for an exemplary model of deployment.

- Look for phased deployments so that benefits of certain technologies and regulatory changes result in cost savings, automations, customer satisfaction, and a positive public image of parking which has been lacking since meter technologies and PARCS were invented. In addition, there should be provision for changing course if expectation is not met in one phase.

The IPMI's Framework on Sustainability is a good start to move forward to the next era of the parking industry revolution. In this chapter, additional elements have been identified and defined. Hopefully, sincere and honest efforts by all participants to keep up pace with technology will bear fruit. Important areas to be explored are less hardware, more software, less or different regulations, ITS with limited resource sharing, and web portal for both capital cost savings and yearly maintenance.

CHAPTER 8

Easing Parking Regulations

INTRODUCTION

The auto industry has changed our livese since its invention, creating an environment of traffic and parking regulations and enforcements for the safety and security of all citizens. The impact of this environment on the upward mobility of inner-city residents has come under question lately. The Department of Justice (DOJ) report on shooting deaths and unrest in several cities in 2014 and 2015, including Ferguson and Baltimore, documents the evidence.[1] We are talking about traffic/parking criminalization leading to discrimination and disparity due to the combination of greed for money and human prejudice. Regulations are required among many things to bring social justice and fair treatment of people. Is it truly happening? Not really. Traffic and parking violations and associated enforcements will be discussed here to show overregulation combined with human judgement and bundling of many social issues may generate criminal charges leading to discrimination and disparity in the society – a leading cause of economic disparity. Once caught, the poor are being trapped in a downward spiral at different federal, state

---

1 Mark Berman and Wesley Lowery, 'The 12 Key Highlights from DOJ's Report Scathing Ferguson', *Washington Post*, March 4, 2015. Eric Lichtblau, 'Justice Department to Track Use of Force by Police Across US', *NYT*, October 13, 2016.

and local levels. Easing regulations and removing human judgment with technologies will be highlighted to bring improvements.

Modern technologies can help solve social woes if there is the political will. Statements like, 'We Want America Back' by political leaders do not serve any useful purpose but evoke a negative connotation of white-only rule. America of the 21st will not be the same as America in the 20th century because of the changed global economy and technological landscape. We need to explore the reasons for increased discrimination and disparity. The inner core of human values is not polarized by political bias – the sooner we realize this the better. America's future does not seem equitable for all – it is not a matter of left or right. Reasons include the ineffectiveness of the Congress, wealth concentration in few hands, the gradual disappearance of the middle class, and the growing gap between the poor and the rich. The auto industry may have contributed to this. To top it, police brutality causes occasional outbursts in inner cities. Events in Ferguson, Baltimore, Bronx, Charleston, and other cities should provoke a real solution not reactionary responses.

There were numerous articles, editorials, and op-eds after the shooting deaths of Mr. Brown of Ferguson, Mr. Gray of Baltimore, and Mr. Garner of Bronx. Two recent op-eds, in the *New York Times* summarize them all.[2] There are numerous recommendations on early education, income equality, housing, health care, minimum wage, free lunch at schools, nutrition, food waste and connection to homelessness, prison population and rehabilitation, etc. There are also promising results from trial projects but the continuous funding commitments, along with policy and strategy changes by leaders at federal, state, and local levels, are missing. Continuity of such uncertainty may be a disaster for our future. Something not widely acknowledged yet in the social mainstream is the impact of traffic and parking regulations on the poor: the high cost of tickets, arbitrary increase of those tickets, losing driver's license for non-payment, and high cost of reinstatement. Can these issues be addressed with modern technologies?

This section explores traffic and parking related social woes as they affect upward mobility for inner-city residents. Traffic and parking

---

2 Orlando Patterson, 'The Real Problem with America's Inner Cities', *NYT*, May 9, 2015. Jason Furman, 'Why Investing in Families Works', *NYT,* May 11, 2015.

violations and criminalization are due to recent over-regulations whose severity could not have been fully assessed before the Ferguson and Baltimore incidents. We are talking about discrimination and disparity – a byproduct of too much criminalization in the transportation (traffic) and the parking industry. Policy makers found it convenient to add other social and unrelated punishments to this criminalization. Greed to make money has grown out of control because of over enforcement with bad human judgment. A self-sustaining traffic and parking business model leading to an infrastructure and a new paradigm without human intervention may be explored from other industries such as telecommunications and airlines. Virtualization of telecommunications industries with intelligent software building blocks has reduced the cost of long-distance voice and data calls. The workload sharing approach with passengers, such as online boarding pass and bag check-ins of the airline industries, has defined a new paradigm for a disrupting online business models. Both of these business models seem to have some relevance for the traffic and the parking industry. Both industries may face some disruption as they move to the next level of sophistication. That is very normal as long as they are crafted to derive the desired results.

There were reasons for the many regulations that exist in the parking industry today, but with the advent of new technology we need to rethink some of the regulations. Thirty-five years' experience of the telecommunications and information technology (IT) service industry have led me believe that that civility must prevail in terms of customer satisfaction and convenience. An exposure to the parking service industry for the last twenty years has given me a different impression. No other service industry draws such negative media attention.

Only in the parking industry are you given a ticket (a criminal charge in some sense) for overuse. Nothing like this happens in any other service industry such as telephone, electricity, airlines, and even a city's water and waste collection. Those service providers may charge you at a higher agreed upon rate for overuse. In the parking industry, some say that is just how current parking regulations work and some say that this is how city's budget is balanced, with extra collection from parking violations, or that is how the city can pay bonuses to the ticket writers or police officers from this ill-gotten fund – close to $18 to $20 billion per year – of uncollected fees boosted by the associated penalties in U.S.

A. Many cities have come up with incentive plans for drivers to pay off their fines with no penalty but it would be much better to change in favor of VIP like parking services.

Regulatory ease is required for long-term goals such as Google's driverless parking, stronger public and private cooperation for optimum resource sharing of mutual benefits, improving construction codes to maintain commercial/residential ratio to parking areas and 'Go Green' evolution strategy with solid CO/CO_2 emission codes in cities including enclosed garages/lots. Let us ask some penetrating questions about the current regulatory paradigm to get some new perspectives.

- Why parking ticket? Replace it with an online account so that violators could be charged to their accounts as in other similar city or utility services at increased rate – pay-as-you-use.

- Why meter-specific parking? Try out meter-independent but hours (or day) dependent parking with a home-printed parking pass.

- Why handicapped decals or other decals? All car owners and their license plate information should be in the database for auto check of special consideration.

- Why take every parking offense to court? Introduce civility with the help of technology so activities are well documented and advance payments are made.

- Why feed the meter with cash for parking? Introduce an 800 call for parking activation/deactivation while sitting in the car so that cash collection can be avoided.

- Why expensive meter, inventory, and field maintenance? Deploy a parking portal to replace meters with marked spaces and online collection.

- Why have idling or circling in the city lot or the garage? Make space reservation and avoid CO_2 emission and traffic congestion.

- Why not divide work responsibilities between the parker and the administration? Follow the model of the airline where you can print out boarding pass and luggage check from home or even in the cellphone.

- Why not increase the mobility and convenience of parkers by extensive use of cellphone (iPhone in particular)? With simple registration and mobile PIN code, the parker has the flexibility to activate or deactivate parking privileges (street parking, garage parking, event parking) from any place, and in any time.

- Why do you need two separate processing systems (street parking and garage parking)? The web portal should be designed to integrate both.

- Why do you need complicated and vendor proprietary Parking Access and Revenue Control System (PARCS)? Simplify it with less labor-intensive Web Portal Applications (with easy audit) so that cellphones and the Internet are integral.

- Why not fold parking services under ITS? Joint planning will save competing resources (private/public cooperation).

- Why not connect all parking resources nationwide like the telephone network and electrical grid? Display the availability of parking space at the entrance to the garage.

- Why have office visits? Reduce the number of employees with online operation, account management, payment processing, auto payment processing, complaint/appeal handling practice, and to deal with other queries.

No one wants to see the traffic courts full of violations nor to use scarce police resources to regulate car parking when modern technology makes this redundant. Regulatory adjustments need to be identified and made.

CRIMINALIZATION ASPECTS OF CURRENT REGULATIONS

- **Traffic Regulations:** Traffic control and associated regulations started with a noble idea of monitoring cars on the road for erratic driving, illegal stopping and parking, collecting parking fees and fines, speeding, and other irregular auto movements. In other words, social harmony on the road and the parking space was the main issue. The process was very manual with mostly crude implementation that was possible with the then available technologies. Social change, human judgment, and over-regulation have changed the landscape, causing discomfort for many of us.[3] The result is an exponential growth of traffic woes. Fortunately, innovative software technology can address many of them to upgrade the system and the process. That is why it has become an imperative task for the industry to have a critical look at traffic and parking regulations and its criminal aspects such as well perceived discrimination, criminalization and disparity as they exist in the industry. Hopefully, a solution to the parking problem could reduce other traffic related discrimination, criminalization, and disparity.

- **Traffic Discrimination:** Was there intentional discrimination by city officials in penalizing traffic violations? Nobody thought so openly before the shooting of Michael Brown, an 18-year-old boy, by police in Ferguson, Missouri, in August 2014. A *New York Times* article[4] clearly states that Ferguson became a symbol but that biases extend to many cities nationally. There are different kinds of traffic violations and parking is one of them. Others are speeding tickets, illegal stops, and traffic lights, etc. The report established facts in three areas with solid statistical data.[5]

Ferguson Police and Officials Discriminate Against Blacks

Figure 8.1 is a sample of biased data collected by the NYT reporter. It is evident from this that the police were biased and the city officials violated the constitutional rights of black residents. Categories of violations are: 1) police and court employees expressed racist views in emails, 2) searching cars owned by blacks without cause at gunpoint, and 3) justifying the use of force with a patently untrue statement. These data could be extrapolated to other cities or even nationally to confirm that the discrimination exists at a level that cannot be ignored anymore. A short-term approach just to address discrimination may not be an optimum solution. We need a long-term solution with an integrated approach.

---

3 See my published papers, listed in the Bibliography to Chapter 1.
4 Campbell Robertson, Shaila Dewan, and Matt Apuzzo, 'Ferguson Became Symbol, But Bias Knows No Border', *NYT*, March 7, 2015.
5 Wilson Andrews, Alicia Desantis, and Josh Keller, 'Report: What is Wrong With the Ferguson Police Department?', *NYT*, March 4, 2015.

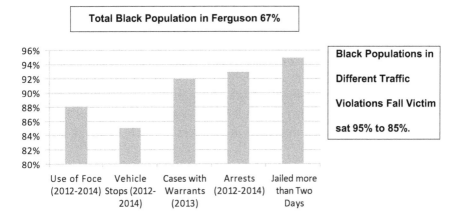

FIGURE 8.1   Higher percentage of black population victims.

Police and Court Officials Focus on Making Money Rather than Ensuring Public Safety

It is a known fact in the parking industry that the city increases its budget funds by generating extra revenues through parking fines, late fees, and collection charges. The extent was not known to be as bad as was cited in the DOJ report. Pressure was put on police and court officials to target offences with higherrevenue generation. A few examples of such pressure were given in the report. The police chief boasted of a bumper month when he collected more than usual by $17,000. There fines continuously increase if you cannot pay once and even jail term will not eliminate the payment. There were also excessive fines above regional averages for minor offenses. Figure 8.2 (copied from the same report) shows the year-by-year steep increase in revenue generations.

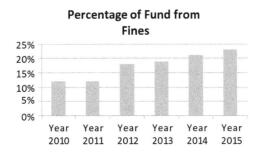

FIGURE 8.2   Revenues generated (%) from fines.

Similar Comparison of Other Missouri Cities like Ferguson

The question was also raised by the DOJ report whether the pattern in Ferguson reflected other Missouri cities of 10,000 people or more. The report concludes that the pattern is similar to other Missouri cities on several measures. The police stopped black drivers about 1.4 times more than their share of the population. The municipal cities collected more than $80 per resident in court fines and fees in 2014 that locked poor black residents in a spiral of debt, jail, and court appearances.

One can summarize the DOJ's investigative report by asking following questions. Once these questions are thought about one can hope for a long-term solution regionally and nationally.

- Is random ticketing unique to Ferguson?

- Should the perpetual cycle of steep fines be the norm for recession-proof revenues?

- Should there be a cap for traffic tickets and court fees?

- Should there be an alternative to jail for people who cannot pay?

- What happens if the defendant is black and the judge is white and there is collusion?

- How to avoid ticket fixing?

- How is the city's budget balanced?

- How was such discrimination kept secret for so long?

There are regulations on each of these points and each is subject to manipulations. According to the DOJ report, these issues reflect discrimination against African-Americans and poor people. In some cities across the nation where there is a large black population 30% to 40% of city's budget is raised from poor people. It sounds very alarming. Can modern technology solve these issues? Probably not. But I will make a case for parking services that will at least address them. It is hoped that some of the prescriptions recommended for parking violations can be extended to other traffic violations across all states and nationally.

Disparity and Driver's License Suspension

How far do you go to penalize auto owners for simple traffic or parking violations? A reporter quoted California motorist who lost his driver's license and job due to non-payment of his escalated fines. 'I'm tired of being broke,' he said. 'And I'm tired of not being able to pay my tickets.' Unfortunately, these escalated fines are supposed to pay city's other budget and bonuses to officers to be harsh to motorists, especially minority motorists who cannot break the cycle to rise above the poverty. In Tennessee, state legislators wanted to raise $20 million a year from such penalties with no consideration of the consequences for the poor people. In California, it is not unusual for $25 fines to escalate to $2,900.[6] In recent years, many states have gone back and forth on these policies amid concerns of not hurting low income people. For example, in Washington State, there were 500 fewer arrests for driving while suspended, saving an estimated 4,500 hours of patrol officers' time, when suspension was stopped. Driver's license suspension for non-moving traffic violations is not trivial. The suspension process became so weird that in some cases people's licenses were suspended even though they did not have one. It most affects the poor who cannot break the cycle due to biased and harsh regulations.[7] Judges may have discretion to break the cycle but seldom use it to show a sympathetic attitude for administrative reforms. Figure 8.3 gives national statistical data on crashes compared to total suspension. The number of crashes does not go down when suspensions go up. In addition, there are complaints about police officers planting false evidence to favor the prosecution. Is it fair? No, and that is why we need to have a serious look at changing the disparity, discrimination, and criminalization process of current traffic regulations to a more liberal form of regulation. Technologies, if properly exploited, could help us to achieve a just society. New parking technologies will be one of such focus in this chapter. As described in the RWPTTP,[8] there will be no hard meter for on-street parking and traffic monitoring at highways will be automated with no human touch for unfair treatment. There will also be no gate or arm for off-street parking. GPS in the smart cellphone or in the car will

---

6 Timothy Williams, 'Disparity Is Seen in California Driver's License Suspensions', *NYT*, April 8, 2015.
7 Shaila Dewan, 'With Driver License Suspensions, a Cycle of Debt', *NYT,* April 14, 2015.
8 Robust Web Parking, truck and transportationPortal (RWPTTP).

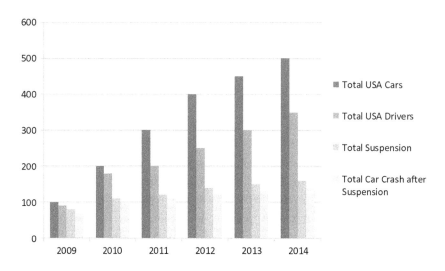

FIGURE 8.3   Crash data even after suspension.

replace the existing proprietary and expensive hardware system. Software will emulate all hardware parking functions in the form of parking apps. The application will automatically identify the vehicle and the owner when it approaches a space of his/her choice. In addition, setting up an account, reservation, and prepayment with no cash transactions can simplify the enforcement process. Once violations are removed from the parking fee collection process the use of judicial resources will be reduced. Unmanned drones can monitor erratic driving and capture a driver's identity automatically. This is one way to consolidate transportation services for convenience to end users.

DECRIMINALIZING PARKING AND OTHER TRAFFIC SERVICES

Auto traffic has always been a hot topic in the business and political community of many cities. There are many stakeholders of the industry including local government and community leaders. One instance is the dissatisfaction expressed in local and national newspapers by numerous potential patrons of downtown cultural and commercial venues because of the punitive and predatory nature of the current traffic regulatory system. A lack of foresight about the rapid growth of downtown and biased parking fees may be blamed for the current state of traffic in the city. Articles have mostly discussed meter and no-meter, fee and no-fee parking, business interests and recreational value, stricter policies and relaxed policies as possible solutions to the problem. Here my focus is the idea of virtual transformation to be independent of human interactions, using technology to solve city's parking woes. No regulations are perfect but complex regulations may not work in the social environment the auto industry has created. Simpler regulation that could be automated with software implementation will free up court traffic and reduce the heavy use of judicial resources. Revolutionary thinking is required to create an efficient, flexible, modern, and customer-friendly system. Technocrats, parking administrators, parking operators, and the city council must work together to bring the desired change – a change that brings customer needs and city needs together with civility, mobility, load sharing, and other cost effective measures. In discussing a technological solution to this problem, I will focus on three themes: harnessing technological progress, updating city regulations, and new revenues with a real sense of customer service.

In the November 2006 issue of *Parking Professional* magazine, I first raised the criminalization aspect of the parking industry in an article, 'Decriminalization of Parking Services – A Paradigm Shift'. Aspects of discrimination and disparity were present but were not highlighted by the author. The latest turmoil in the media about the city of Ferguson and has given us reason to rethink the way traffic tickets are managed by different cities across the nation.[9] It has been revealed and proven again and again that traffic tickets may cause some discrimination and disparity in addition to criminalization. This may be attributed to human handling and out of touch regulations. Virtualization of traffic tickets processing from start to finish may be a possibility. The process should be auto-mated to be totally anonymous with minimal human interaction.

I will start with parking service regulations. Decriminalization of spaces where parking is allowed is under consideration and discussion here. These on-street parking regime may not be needed once the RWPTTP premise is accepted:

1) Meter to accept payments

2) Cash collections by meter maid

---

9 John Eligon, 'Ferguson City Manager Cited in Justice Department Report Resigns', *NYT*, March 10, 2015.

3) Fines for time violations

4) Restricting hours for parking

5) Wheel-locked procedure or towing for legal parking spaces

These off-street parking aspects also may not be needed:

1) Restricting access to garages by arms and gates

2) Attendants at the gate

3) Stop, pay and go at the entry/exit

4) Auto payment machine for ticket payment

Instead, we need the following modernizing system in allowed parking spaces:

1) Parking spaces to be declared national resources, to preserve prime lands

2) Workload sharing with patrons like airlines

3) Soft Meter Application (SMA™) for both on-street and off-street parking

4) Making cities smarter with WiFi connections

5) Connecting all mobile devices in a network for an efficient integrated parking and transportation system

6) Cost reduction by technology and more revenues for new services instead of arbitrary rate increases

7) Similar rates for day and night parking

8) Instant change of rates for special events if approved

9) Advance collection for guaranteed and reserved spaces like airlines

10) Integration of toll collection, parking collection, and other traffic violations for a better transport system

11) Automatic step-up rates for any overstay in an approved space

12) Time stamping logs of all events for review as and when required

13) Online resolution of complaints with audio-video clips

14) Introduction of self-driving and self-parking cars

15) Future drone parking policy overview

My contention is that technologies are available now to make revolutionary changes necessary to introduce a new parking service paradigm. The key to a successful parking program will be the flexible integration of the multiple ways in which potential patrons interface with information: web technology, the cellphone, hand-held devices, interactive voice response systems (IVR), geographical information systems (GIS), color-coded bar code readers, license plate readers (LPR), scanning of permits, RFID tag, PDA, WiFi, etc. The SMA™ preview shows one particular implementation that integrates these systems in a way that maximizes benefits for both the municipality and the customer. This particular system can be further customized and improved as technology progresses. In creating a uniform parking application, the problem is that the array of services is very complex – street parking, airport parking, event parking, sporting events parking, garage parking, etc. The needs of customers in short-term versus long-term parking, daytime versus night time parking, can be diverse as well. We can develop a unique and robust application to address all problems associated with each particular parking scenario. To demonstrate, I will present a situation involving a potential patron of downtown amenities, the current way that his parking needs are met, and the way that his/her parking needs would be met in the decriminalized and, hopefully, the discrimination-free system.

Step1 – The Scenario

You are driving from City A to City B for a business meeting. Your meeting is at a hotel in a busy street, downtown. Your meeting starts at 11 a.m. and finishes at 1 p.m. You come with your wife and would like to attend an evening concert at the City Memorial Auditorium, South Street, before returning home late that evening. While you are in the meeting your wife drives around downtown to see the sights and to go shopping. How could you make your driving and parking a pleasant experience without unpleasant regulatory encounters?

Step 2 – Current Approach

You go to Mapquest for directions from City A to City B, and print them out before you start your car. You find South Street and see your hotel. You then look for a public garage to park near the hotel, assuming the hotel charges $3 more than the available public garage. You may have to circle a couple of times around the hotel block to find a suitable place, increasing the congestion and pollution in the city. You become frustrated as it looks like you may be late for your meeting. When you finally reach your meeting, your wife goes out with the car and parks on the street to buy some souvenirs. She is one minute late to come back after shopping. To her dismay, she gets a parking ticket costing her $10. She either pays the ticket or fights the case (through a process that may take several court dates for resolution, and may cost several hundred dollars in legal fees). Later, you buy another parking permit for the concert in the evening but drive around (congesting roads during high traffic time and adding air pollution to the environment) to find a suitable place close to Memorial Auditorium. All in all, this is a clumsy process.

Step 3 – An Intermediary Approach

While the best option would be for all parking to be free, the second-best option is to implement a system that properly exploits technology in combination with flexible new parking regulations. This option would replace the traditional, inflexible, unattractive, hard-line approach to parking regulation. Potential features and functions of the new system would include:

1. Prior arrangement of a payment method with an appropriate authority as is done with other public utility commissions (telephone, water, electricity, etc.) or purchasing a parking credit (registration process using a website).

2. Making reservation of a parking space (prepayment for a guaranteed space) using an Interactive Voice Response (IVR) system.

3. Agreeing on standards among parking authorities or third parties to share information and payments (nationwide Application Service Provider – ASP).

4. Step-up rates for longer occupation of resources and debiting the account.

5. Making drivers responsible for their part of work.

6. Exploration of cellphones for instant payment.

In fact, cellphone companies already have many of these features and may be able to support municipalities in their development of this infrastructure. This type of billing makes accepting parking payments easy and time sensitive with potential for flat billing and step-up billing for overstay – it's like collecting fines online instantaneously rather than going through the overtaxed judicial system. The parking authority can also implement this sort of system themselves once they build up their own IT infrastructure through the so-called WiFi capable smart city.

Instead of ticketing for overstay as is done currently, the following simple procedure would be a possibility. You register with the parking authority for parking credits; once your credit is established for a certain parking limit you are free to park anywhere and anytime with the help of interactive voice response (IVR), as is done for telephone, water, and electrical usage. You may be given a PIN or special code for such privileges. In a street parking scenario with a space number, you park in a space first and then call an 800 number using your cellphone to activate a parking privilege. After you provide a PIN code, the system will recognize your parking privilege and the IVR will guide you to set the parking start time, end time, rate, and total charges for that particular space and for the duration you choose. It will even let you deactivate your privilege when you leave even if it is before the time you initially chose. Once the privilege is deactivated before the initial period your charge will be adjusted based on your duration and according to the city regulations. No physical meter or coin is required for such procedures. Physical meters are replaced with a smart meter application (SMA™) – a centralized software package for all spaces (meter or non-meter) – saving a city millions in capital investment, maintenance of meters, inventory of parts, coin collection, etc. Instead, you enhance your IT infrastructure to include parking services along with other services.

You need a second procedure to handle parking time violators. I do not recommend ticketing as is done now because it is harmful for a city's image. Instead, two options are possible for such violations. The parker voluntarily extends his time using the Internet or cellphone within the limit of the city regulation. If the parker does not do so, the city can extend duration and charge more for the extended duration as permitted by the city or step-up rates equivalent to ticketing. To rationalize and to document such actions, the system can send a warning message first for an action as permitted by regulations and execute the action based on the warning if there is no response, using short message service (SMS) via a cellphone. All interactions are time stamped. SMSs or emails are exchanged as an official record to avoid disputes with parkers. Revenues will be collected graciously and electronically with no complex court proceedings for simple time violations. This may be the simple building block of parking decriminalization. More complex modules can be developed as we gain experience.

Coming back to you and your wife's trip to City B, while at Mapquest, you click on the City B parking website for all details of parking garage, parking space availability in the garage and street, street parking area, rate, distance from the hotel, etc. While at the website you now select a garage, level or even space of your choice and make a reservation for a day or so and get a printout of your reservation to use for parking in City B. If you are willing to pay more, the city may extend your privilege to park at any of the garages around the city including the garage close to the City B Auditorium for the evening concert. In addition, you can also get another permit or privilege for your wife to use for on-street parking. Or, for getting a one-day garage-parking permit, some cities may make street parking free for you if you display your permit at the window. In fact, you can solve all your parking needs with one click – garage parking, street parking, and concert parking. You can now go directly to the parking garage of your choice, without circling around, and present your parking permit. Your wife can park anywhere on the street to shop and buy souvenirs without worrying about parking tickets or a court fight. If you share your license plate information or type of car and color, the parking authority will recognize your license plate for the royal treatment. If you violate any traffic rule you may be warned of consequences with a time limit for corrective action before action is taken against you.

In summary, the decriminalization process involves evolutionary changes in regulations to include the following:

- Registration with the city/municipality for an account

- An account privilege to park anytime, anywhere

- Reservation of a parking space for close proximity privilege
- Documentation of all transactions/correspondences for online viewing by all stakeholders
- Introduction of PIN for parker's security identity
- Identification of parker with License Plate Reader (LPR)
- Use of cellphone for parking mobility
- Step-up rates for overtime usage
- Elimination of ticketing for parking fine collections
- Keeping all records for arbitration instead of court proceedings

The integration of the software technology into the parking system can also reduce the need for a cumbersome, cash-oriented physical infrastructure: eliminate parking meters with the smart meter application (SMA™), replace time violations with step-up fees, and replace traffic tickets with an online warning and instant collection of violation fees. In contrast to the cash system, all activities are recorded to improve fairness, as is done by banks, electrical companies, telephone companies, and by municipalities for water bill collections. Furthermore this system eliminates those familiar altercations between the driver and the ticket-issuing officer if they happen to meet 30 seconds after the meter expires.

Further benefits of this paradigm-shift are as follows. For the parker, the benefits are …

- *Convenience.* The driver can avoid endlessly circling around a parking lot looking for a good space. You know where you are parking even before you start the car.
- *Saving time.* Not only does the driver save time in getting into the parking lot and driving straight to the space, the premium spot means the walk into the airport terminal (or sports venue or whatever) is shorter as well.
- *Flexibility.* The driver is able to make new reservations and change existing ones while on the move.

- *Improved information and peace of mind.* By having access to information like directions to parking lot and distance from the lot to a given destination (a terminal, a museum, the arena box office, etc.), the driver is better able to focus on more important issues.

For the parking authority, benefits in addition to a better rate of returns (ROI) are:

- *Increased revenue per space.* The parking authority is able to generate more revenue from the same number of parking spaces by stratifying its customer base.

- *Better customer service.* The parking authority can improve customer service to passengers, fans, and other end users.

- *Improved traffic flow in parking lots.* By providing dedicated parking areas for premium customers, the parking authority can reduce the number of cars using a given entrance or exit, thereby improving overall traffic flow in the lot.

- *Advanced fee collection.* The parking authority receives both the base parking fee and the reservation premium before the parking event.

- *Reduced capital investment.* More spaces can be added to the public parking inventory, to increase the revenue base as and when desired, with small software changes, replacing capital requirements for buying meters and other parking equipment.

- *Improved cash management.* By reducing the amount of cash handled by employees, the parking authority has better control over its revenues.

- *Better cost control.* Automating such tasks as enforcement and ticketing reduces overall cost structure and requires less manpower and less time.

- *Improved security.* The parking authority is able to improve overall security through identification of customer license plates, authentication and authorization with digital signature, and LPD/OCR (camera to computer License Plate Detection/Optical Character Reader) integration for total automation of the parking process.

- *Improved information about customers.* The parking authority can collect valuable personal and vehicle information about customers.

Of course, no miracle is possible in the short term. Drivers have to be educated, employees have to be trained, and technology has to be trialed for maturity. The potential looks great if we work together to move in a positive and constructive direction. This 21st-century approach to parking can decrease congestion, decrease pollution, and create a warm and inviting atmosphere for patrons of downtown City B. By all means, municipalities must investigate all of the possible solutions available to them including the complete reorientation of the current configuration. Halfway measures limited by outdated regulations and pre-Internet thinking, however, will only keep City B from becoming everything that it has the potential to be. With people increasingly dependent on cellphones, the Internet, hand-held devices, and other personal electronics, an intelligently designed parking system that integrates these technologies in a customer-friendly way is the only way to be successful. This narrative has been articulated for the forward-looking mayor and councilmen to understand the problem and its solution now and vision the future. This may also be one way to influence the city council to make smart and visionary decisions about future parking. I am convinced that once a city/municipality starts this paradigm shift, private companies will orient themselves to be competitive and also bring the desired changes in the industry.

Bluetooth Beacon (BT) – Better Option for On-Street and Off-Street Parking Integration

There are many newer technologies on the horizon to improve the option described above. They are Bluetooth (BT), GPS, and RFID tag. Motorola has come with a BT device called Gamble to interact with smartphones and send vehicle identification to a central server for different application development. On-street and off-street car parking described below could be one such application. Modern cars fitted with latest GPS will also fall under this category. The same application will bring equilibrium to the rate of on-street and off-street parking. This will solve the disparity that has been created in the industry.[10] In fact, an elegant implementation will be to follow I-540 toll collection methodology – better than stop and go as is done in EZ-Pass system. Each driver can register his parking privilege and in exchange he/she is given

---

10 Shoup, *The High Cost of Free Parking*.

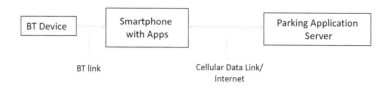

FIGURE 8.4 Configuring BT for parking application.

a RFID tag for parking or toll. In other words, the same tag can integrate both toll and parking fee collections. It will be a low-cost option for the administration because, first, there will be no meter and, second, the customer or the third party vendor will share some of the cost. It will bring a new friendlier user experience. The process will include the following:

- Making the smartphone smarter, to accurately correct the GPS location of the space with a new icon

- Full use of integrated GPS in the vehicle without BT beacon connection

- A server software to interact with the smarter phone for parking activation and deactivation

Steps of On-Street Parking Transactions

After location correction interacting with the server software one can initiate on-street parking using smartphone icons.

There is a common thread in the traffic and parking services – virtualization of traffic and parking services. You are billed with a time stamp – 99.9 % accuracy. Other than the initial technological glitch no major complaint about the toll collection system has surfaced. You may face some penalties if you do not pay on time. Establishing a credit system with an account and a tag can avoid that too. LPR can be extended to both the traffic and the parking industries. Other technologies beyond LPR that could alleviate many woes are: intelligent software for robust web parking truck and transportation portal (RWPTTP), online reservation of parking space like airline seats, geo positioning software (GPS), Google maps, Bluetooth low energy (BLE) device, Radio Frequency Identifier (RFID), and most importantly drone technology. The drone technology with high

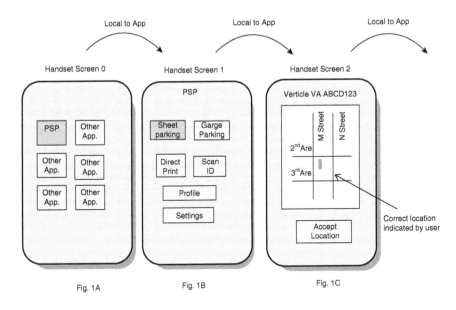

Street Parking Screens: Start Parking

FIGURE 8.5 Smartphone use activating parking application (street parking screens for start of parking).

precision cameras could be useful in two areas – core traffic violations such as speeding, irregular movement, and illegal lane crossing; and monitoring and enforcing parking violations in open airport lots, hospitals, and government key installations without gates. Drone technology could bring disruption and revolution in both industries once enforcement is automated. Garages will need smart wiring with WiFi and BLE technologies to automate the operation so that human judgment is avoided. Intelligent software, BLE with smartphone, and GPS embedded in smart cars can replace expensive street meters, their maintenance and daily field operation such as cash collections. Software blocks can emulate the current hardware functions for better customer interface, new service for increased revenues and efficient enforcement. In addition, the same software can be configured to cover street meter and garage parking applications. In fact, we are not far from all auto owners having one account for parking, toll, and traffic violations. We need to stop crimes and civil act violations. Fighting poverty is also important to decrease crime.

Many solutions are possible but people's perceptions can be twisted by politicians, exploiting a racial reaction in the face of growing social woes, unemployment, and lack of job opportunities for certain ethnic groups. We need to explore all reasons behind such increasing discrimination and disparity. While at it, we need to explore other social woes that impact an upward mobility in inner cities but are not talked about much: food waste and connection to homelessness, prison population and rehabilitation. Here I am attempting to shed some light on traffic and parking regulations that I have been researching for some time. After the Ferguson and Baltimore incidents many social pundits are realizing the importance of addressing those issues as part of mainstream problems. Criminalization leads to discrimination and disparity and human greed to make money has grown out of control because of outdated regulations.

How can an elegant solution avoid unnecessary bundling of all social woes under the traffic or parking acts? Simplify regulation, focusing on industry-specific guidelines and use modern technology to automate the process to avoid the human element. The objective of all traffic and parking regulations and policies was to deter violations, not make money or solve unrelated social woes. Are they working? May be for the rich because they can pay fees, fines, penalties, and associated reinstatement remedies. But escalated amounts hurt inner-city residents the most not only in the U.S.A. but in the world. In Scandinavian countries, these fines and penalties are proportional to income. One example is Reima Kuisla, a Finnish businessman, who was caught going 65 miles per hour in a 50 miles zone in his home country. The police pinged a federal taxpayer database to determine his income, consulted their handbook, and arrived at the amount that he was required to pay: €54,000 or US$103,000. Should we explore such options or look for a better technological solution? May be a combination of all with a right set of new policies.

To overhaul the traffic and parking current infrastructure, I suggest we look at three pieces of research. First is the 2013 report by the American Association of Motor Vehicle Administration (AAMVA), based on analysis of data from eight states.[11]

- All fifty states, the District of Columbia, and many Canadian provinces have provisions to bundle traffic and non-traffic related suspensions.

---

11 American Association of Motor Vehicle Administrators (AAMVA) Report, February 2013.

- Bundling many social issues with traffic violations defeat the intended purposes and hurt the poor the most.

- Bundling of these social issues varies from non-payment of child support, carrying firearms, etc. to small theft, state to state.

- Length of suspension also varies for same crime, state to state, starting from a few months to an indefinite period.

- Non-payment of child support tops the list of all non-traffic offences.

- Repealing regulations for suspending driver licenses for non-traffic violations will increase traffic violation effectiveness.

- No substantial evidence was found that suspended drivers fully comply with court orders or legislative orders.

- Driving and possessing a valid driver license provides upward mobility for poor people.

- 75% of suspended drivers drive in spite of suspension, defeating the sole purpose of its objectives of road safety, putting tremendous pressures on law enforcement resources.

- Drivers suspended for traffic offences are more dangerous than other suspended drivers and appropriate action is appropriate against dangerous drivers.

- By 2014 estimates, there are over 84 million suspended drivers in the U.S.A. and the number is rising.

- 39% of drivers suspensions are non-traffic related and people cannot afford to pay the layers of penalties imposed during different stages of the suspension.

- Suspending drivers for social non-conformity dilutes the severity of the traffic violations because many of them drive anyway and receive multiple suspensions.

- In general, the local community feels that drivers who keep their licenses will keep their jobs in stable employment and this will improve the local economy.

- Public traffic safety is not helped by suspending licenses to encourage social behavioral change (e.g. for non-payment of child support).

- Each year new driving legislation is introduced which has no relationship to operating a motor vehicle.

- Drivers suspended for traffic safety reasons cause three times more crashes than those for non-traffic suspension and it is society's responsibility to keep them from operating a vehicle.

- Drivers who cannot pay for the first suspension for non-traffic violations are trapped by subsequent penalties, almost ruining his or his family's life, but not bringing any social benefit.

Eliminating the 39% from the traffic citations which have no relation to motor vehicles will alleviate clogged court dockets. It will also free limited resources to fight other emergencies. It will also help 39% suspended drivers to retain driving privileges to earn a living and contribute to the economy. Significant savings are possible in the DMV budget if DMV is not responsible for processing social non-conformance can concentrate on its core business. What is the use of penalties if they are not collectable? Moreover, one needs to question the logic behind such policies if those who cannot pay are mostly teenagers or the poor. Many studies have shown the negative impact (losing job, not finding a jobs, or decreasing income) of suspending driver license for those who are productive but not dangerous on the road. Social compliance is important for traffic violations but not by bundling. There may be other options such as bank or salary garnishing for non-social compliance so that the daily ability to work or economic prosperity is not jeopardized. For traffic safety, technology options should be explored.

What should be the objective of the driver suspension in future?

- Separation of traffic and non-traffic regulations and policies for better customer service.

- Automate traffic violation monitoring with less human touch.

- Bank/wage deductions for nonpayment for some suspensions.

- Amnesty and diversion programs in lieu of suspension.

- Ignition interlock for all cars, with proper security and safety in place against hacking and cheating.

The next piece of research to consider is my November 2006 paper, 'Decriminalizing the Parking Services – A Paradigm Shift', in *Parking Professional*. Findings of the paper are:

- City parking regulations are punitive and predatory.

- Parking services are not customer friendly or at par with other city and public services such as water, trash collection, telephone, and electrical services.

- There is no national standard for making parking services uniform.

- There is no public and private cooperation for the best utilization of parking spaces.

- Parking rates and penalties are raised arbitrarily to balance other city budgets, hurting poor people.

- Too many court resources or collection agencies are used for chasing drivers who cannot pay.

- Parking news always makes headlines due to the general dissatisfaction of its patrons and business communities under the current management policies.

- Revolutionary changes are possible with modern technologies (smartphones and smart cities – WiFi) to reduce capital investments, operation costs, and hardware devices in the field

- Online and real-time web operation of all parking services is possible to eliminate parking gates and meters on the street and simplify enforcement.

- City council politics are too complex for fairer enforcement to collect the right revenue or even change outdated regulations for new user-friendly regulations.

- Multiple payment options (credit/debit card, attendants, pay machine), increase the cost of payment collection.

- Hardware devices are no longer required in the field for parking services as they are sophisticated but expensive and hard to maintain regularly.

What should be the goal for the parking industry in future?

- Auto drivers who subscribe to many other city and government services must be trusted and treated as customer deserving highest respect.

- In exchange, drivers must behave as responsible citizens, paying their bills on time and receiving top-quality service as they do from many utility companies.

- Eliminate parking hardware infrastructure.

- Eliminate hardware infrastructure in favor of software infrastructures to introduce online services, reducing costs, as has been done in long-distance telephone services (vonage, skype, etc.).

- Like other city services all drivers should have an account and credits to enjoy parking privileges anytime and anywhere with a smartphone app.

- Single payment option is needed such as invoice and payment, mobile payment, Bitcoin payment.

- Make a reservation for a meter space or garage space through the same web services.

- Share the administrative workload of parking services, similar to airline services.

- Introduce step-up rates instead of tickets and penalties for overuse as the cellphone industries have perfected.

- Keep real-time electronic records of all logs and transactions for enforcements and dispute mediations.

- Increase revenues with customer services such as convenience, time saving, proximity parking, mobility, and flexibility.

- Advance fee collections for better cash management and cost control.

- Enhanced security: knowing who is parking especially in airports, sea ports, and critical government installations.

- Availability of DMV database with vehicle identification number (VIN) and ownership online to detect stolen and hijacked cars automatically.

- Improve traffic flow to reduce GHG emissions and traffic congestion, with a guaranteed online service.

- Automate enforcements with less human touch for dispute handling while introducing drone technology to monitor parking violations.

- Define a parking benefit district (PBD) – a district that uses dynamic pricing techniques and modern technology to make money with free bus services to stores while parking in one place.

This brings us to the third piece of research: the ignition interlock. Auto technologies are going through many changes with new inventions such as electrical cars, solar cars with long battery life, driverless driving and parking. One such invention is the device that checks drivers' breath for alcohol level before letting the vehicle engine start – the sober driver has to be authenticated. The device must also be cheat proof. What is the use of it?

- To stop drunk driving which kills innocent people.

- To stop repeat offensders who refuse to take a blood alcohol breath test.

- To conform to different levels of alcohol for impaired driving as dictated by different state laws (0.04 to 0.08).

- To remove human subjectivity in issuing impaired traffic tickets.

- Suspension of driver's license is not effective because drivers drink and drive even though they know it is not safe.

- Drivers with one DWI conviction are not less dangerous than repeat offenders.

- Twenty-six states use this interlock technology after the first DWI offense and thus reduce alcohol related deaths and arrests – with less burden for the court docket.

We are at the early stage of ignition interlocks,[12] but much work has also been done to deter drunk driving. Preliminary concepts have gone through many iterations to be a permanent feature in the auto industry. Once accepted it will remove the drunk driving hazard. The ignition interlock device goes by different names such as Ignition Interlock Device (IID) or Breath Alcohol Ignition Interlock Device (BAIID).[13] The idea was first conceived in 1969 by Borg Warner Inc. It was enhanced in 1981 by a New Jersey student and then standardized in 1980 for widespread car use. Early devices had semiconductor-based alcohol sensors and did not hold calibration well. In addition, they were sensitive to altitude variation and reacted positively to non-alcoholic sources. In 1990, the industry perfected the device with fuel cell sensors. The idea behind IID or BAIID is to interrupt the ignition signals to the starter until the ignition is satisfied with the alcohol levels from the breath sample, according to the state the car is registered in. To prevent cheating by someone else, the driver needs to pass another test at random several minutes after the start. Otherwise, the car will initiate alarms such as lights flashing or horn honking. The device is installed at the cost of the owner after the first or repeat offenses. Generally, the alcohol level allowed goes down after the first offense, making it difficult for the drunk driver to pass the test.

The current 'Ignition Interlocks' system has several weak links:

• It works as an add-on not totally integrated with the ignition system.

• It can be manipulated by a second person.

• The cost is not well balanced.

• It is not automated, with additional security for owner operation only.

Foolproof Ignition Interlock Device

The perfection of the technology is required to make it an integral part as standard feature of the future vehicles. The following capabilities will make it very useful device for a permanent option:

---

12 Bruce Siceloff, 'Push Underway to Require Ignition Interlocks for All DWI Offenders', *News & Observer*, June 30, 2015.
13 Wikipedia, updated July 2015.

- **Preventing cheating** – It means owner-specific driving only or designated authorized driver with proper safety such as user ID and password at one level. The second level of protection relates Vehicle Identification Number (VIN) to owner picture or finger-prints or any other biometrics. If cheating occurs in the form of sober person blowing the device there has to be constant and regular monitoring of driver's identity.

- **Hacking**[14] – Some results by two well-respected researchers (Charlie Miler and Chris Valasek) on car hacking were published in the *New York Times* on July 21, 2015. Traditional cars, with all mechanical parts, are not subject to hacking. Modern cars use embedded electronics and smart software for their operation, such as remote key, keyless ignition with push button, soft combination keys to lock the car, cruise control, driverless driving and parking. And now come the IID and the BAIID devices. There are many stakeholders in modern car manufacturing and marketing – owners, insurance companies, electronic suppliers, and software writers. The responsibilities of each must be clearly defined for the safety and security for modern-day cars. The findings of these researchers forced Fiat Chrysler to recall 1.4 million vehicles to fix a hacking issue at the request of the National Highway Traffic Safety Administration (NHTSA). Luckily, the finding was in the basement lab and no road mishaps had yet been reported.[15]

- **Programmable to different blood alcohol concentration (BAC)** – This device should slowly be integrated with modern cars' standard option like GPS, google maps, smartphone, and auto direction. There should be discounted premiums for insurance for those not convicted for impaired driving, at the cost of those who are convicted. Programmability will assist in calibrating the device to meet each state-specific blood level concentration, either generally or after each conviction. Per incident installation may be expensive and may not have a national standard. Mothers Against Drunk Driving (MADD) has been advocating the

---

14 Nicole Perlroth, 'Security Researchers Find a Way to Hack Cars', *NYT*, July 21, 2015.
15 Aaron M. Kessler, 'Fiat Chrysler Recall 1.4 Million Vehicles to Fix Hacking Issue', *NYT*, July 24, 2015.

installation of such a device in all cars as standard equipment, like in Japan, Canada, and Sweden.

- **Auto Identification of the owner and the driver** – Owners and authorized drivers must be identified once the alcohol level is detected and before the vehicle is allowed on the road.

- **Keeping all records of activities, 'logs'** – This is required for periodic reviews by the authorities for violations. If violations are detected additional sanctions may be imposed.

OPPOSING FORCES FOR THE CHANGE

Vested groups who make money from the current system or have stakes in the current infrastructure must be discouraged. The legal system will possibly oppose it because the system is a cash earning scheme. Current vendors, providing different hardware devices, make the system very sophisticated but very expensive. Imagine the number of tickets issued and to fight the case the average charge by an attorney is over $200. The attorney pleads two ways – lowering the crime and associate penalty or acquittal without going to the trial. Let us analyze some data nationwide to get the right perspective for developing strategies and policies in two main areas: traffic violations under DWI and parking violations. Driving while impaired under the influence of alcohol is a serious hazard on national highways. Traditionally, drivers suspected of DWI are asked to undertake a breathalyzing test by the enforcement officers. It is illegal to operate an automobile with blood alcohol concentrations (BAC) of 0.08 or above in all U.S. states and Canada. The BAC level is determined by the number of drinks consumed. A drink is defined as different amounts of beer (12 fl oz), wine (5 fl oz), and hard liquor (1.5 fl oz). Males and females will show different levels of BAC depending on body weight after the same number of drinks. The BAC level also depends on other medications each individual takes due to chemical interactions. Each state has its own DWI regulations under the umbrella of Federal General Guidelines. Different variations of the following regulations are applicable for almost all states at different stages of convictions:

- Penalty of high BAC

- Driving license suspension

- Limited driving privileges during suspension

- Ignition interlocks device installation

- Vehicle license plate sanctions

- Alcohol exclusion limiting treatment

- Meeting federal requirements such as:

    ○ open container

    ○ repeat offender

It is hoped that bio-science is making progress to integrate automatic detection of BACs of the owner or the designated driver by the owner. Such automation will prevent drunk drivers from being on the road and technology may also automate future enforcement of such policies.

CHAPTER 9

Optimum Land Use and Pay-As-You-Use Smart Transportation

INTRODUCTION

Let me start the chapter with a story. When I was traveling to Philadelphia for my son's graduation from University of Pennsylvania Medical School, I ran out of cash for the toll in spite of having an electronic wallet and mobile e-commerce capable cellphone. I was annoyed at my wife's reproach for not carrying enough cash and also at the attendant telling me there would be a $25 invoice by mail for a $1 toll. Would you call this civility or good customer service?

My anger had dissipated a few years later when I saw the introduction of a high-speed camera over the highway bypass to capture license plate (so called LPR) for auto identification and also for auto invoicing – a technological improvement for 'No Stop & Go' implementation. No appropriation of land is needed for highway ramps to collect tolls. The NC I-540 toll service is such an example. Yes, the new technology implementation was expensive but the benefits outweigh the costs. Labor costs are minimized and customer service with convenience has improved. The payoff period was just a few years because the huge infrastructure has been simplified.

Similarly, we need to change the current parking paradigm. Not only that, city planners need to coordinate parking and transportation services together to minimize overbuild of parking facilities or eliminate them where necessary. The USA map shown in Figure 9.1 illustrates the vast land available to be used for different applications such as housing and non-housing developments, parking included. Table 9.1 reviews land usage statistics by 15 different categories.

Figure 9.2 shows an underutilized space in an open parking lot. If the vehicle growth continues, parking spaces may occupy more land than housing over the next several decades. During busy seasons, due to high occupancy, payment collection becomes a challenge as well. If you compare item (8) and item (13) of Table 9.1, almost 79% of useful land may be occupied by parking spaces. It may be less if total parking spaces are less. The percentage may be more accurate for urban land than suburban land. In addition, 15 to 16 million vehicles are being sold every year, adding more land for parking especially in the city. It is again postulated that the concentration of parking to housing ratio is higher than 79% in big cities. IBM, as part of their 'Go Green' research, produced the following data as part of their conclusions:

1. In LA business district (5 square miles), driving around in search of parking in one year generates thirty-eight equivalent trips around the world (over 2 million miles). It burned 47,000 gallons of gas, and emitted 730 tons of carbon dioxide. It takes over 10,000 years for CO2 come out of the atmosphere

2. Congestion of roadways costs over $78 billion business loss annually. It also causes 4.2 billion lost productive person hours and 2.9 billion gallons of wasted gas.

3. Driving alone causes 44% of CO2 emissions. Separate data (CO2) of cruising for a parking space only are not known. Construction of parking spaces or garages also produces CO2 emissions. Run-off to drainage of these spaces pollutes water as well.

We are heading to an environmental disaster. Generations X and Y are not happy with the status quo. We need a common sense approach, an 'Affordable Parking Act' to exploit modern technology.

FIGURE 9.1   US map (Alaska is not shown) illustrating the vast land available for different applications, parking included.

OPTIMIZATION APPROACH FOR A REMEDY

Naturally, staying on course does not sound to be a good strategy because the situation will go from bad to worse. There are many corrective measures one can pursue. The most neutral one is not to build any more parking spaces. The most adverse one is to demolish all parking spaces to rescue ourselves from the manmade disaster. Neither is a workable approach but we can do better. We need to follow the motto, 'Do not start the car if you do not know where to park'. In other words, we need to approach the whole problem with the idea of smart parking or the zero cost of civilized parking. Let us now elaborate 'smart parking' or 'the zero cost of civilized parking'. A transportation service element should be added, call it smart transportation service. This out of box approach will have two components so that both parking and transportation can blend together: hard change and soft change.

TABLE 9.1 Land Use Statistics by Occupancy

Total Land:	$3.5 \times 10^{e+6}$ square miles
	(1)
Forest:	28.8%
	(2)
Urban:	2.6%
	(3)
Cropland (suburban):	19.5%
	(4)
Special Use:	13.1%
	(5)
Each parking space:	270 square feet (average)
	(6)
Number of parking spaces:	1 billion
	(7)
Total land used by parking:	270 billion square feet
	(8)
Total housing units:	137 million
	(9)
Average size of a housing unit:	2500 square feet
	(10)
Urban housing units:	3%
	(11)
Suburban housing units:	97%
	(12)
Total land used by housing:	137×2500 million square feet = 342 billion square feet
	(13)
Total number of vehicles:	300 million
	(14)
Parking spaces occupied:	200 million (at any one time)
	(15)

Source: Compiled from a report issued in September 2012 by U.S. Dept. of Commerce, Economics & Statistics Administration, U.S. Census Bureau.

FIGURE 9.2   Aerial view of an empty parking lot.

Transportation Infrastructure Change

We need a backup transportation infrastructure to the current road infrastructure. Why? Roads have become saturated over the years. Technology has already been exhausted to improve it for single vehicle transportation or even plane routes under 500 miles are congested. We need a mass transportation system as a backup system where people can move more freely in large number but not destroy the planet. Two such systems have been identified and are being tested for mass deployment. They are light rail transportation (LRT)[1] as a regional system and high speed rail transportation (HSRT)[2] connecting LRTs. The inherent advantages of rail technology for urban public transportation – high speed, capacity, comfort, reliability, potential for automation, etc. – are well known.

---

1 Mark Marcoplos and Penny Rich, *Light Rail will Benefit All of Orange County*. Orange County, LA: Board of County Commission, February 16, 2019.
2 Peter Stone, 'The Faster Track: Should we Build a High Speed Rail System?', *National Journal*, Fall 1992.

LRT is a technology used for public, and mass rapid transit at a speed of 125 to 200 miles per hour. Such technology is commonly known as a metro or subway, going through many changes since its inception in London 1863.[3] Traditionally, all trains were drawn by steam engines. Now, LRT uses electric power and is built to run as multiple units. Most LRTs ride on steel wheels as in a conventional railway, although some may use different methods. Voltage used may range from 600 to 750 volts using two live rails, one positive and one negative. The practice of sending power through rails on the ground helps routes with limited overhead space. The use of overhead wires allows higher power supply voltages to be used. A practical example of eighteen miles of LRT is being built connecting two large employers (UNC and Duke medical centers) in North Carolina, with nineteen rail stations in between them. The costs range from $2.5 billion to $3.3 billion. Soft benefits are: 20,000 local jobs, $175 million in annual state and local revenues, employment diversity, and lower emissions keeping our cars off the roads. Hard benefits are: transportation hubs, bus or plane routes, relocation of parking spaces from the busy city to suburban areas for autonomous rides, affordable housing along the rail lines, and reduced vehicle ownership[4]. When such LRTs are extended to neighboring cities via rural areas, it will help rural economic development for upward mobility. While relocating parking spaces, they will be calculated on an optimum basis under the environment of reduced car ownership and low housing development across the route. Additionally, the influx of new residents to the city, settling along the rail system, will reduce the strain on our existing roadways and peripheral systems.

HSRTs are wheel-less vehicles with the capacity for much greater speeds than conventional trains. There are two competing technologies achieving more than 300 miles per hour connecting LRT hubs. They are 'Repulsive' and 'Maglev'. HSRTs can cut travel time and save energy once deployed beyond the LRT limit (which may be fifty miles). LRTs will be the feeder lines for HSRTs. Initially, Japan and Germany capitalized on HRTs with early deployment, though the

---

3 Wikipedia.
4 NC State Budget Proposal Could Impact Planned Transit Project – GO Triangle Home News, RTP, May 2018.

technology was invented in USA. Hyperloop is the latest technology of HSRT developed by Tesla and SpaceX which can reach a speed of 750 miles per hour. These technologies can alleviate mounting congestion on interstate highways and dense airport routes of close to 500 miles and 6 million annual passengers.[5] According to the Federal Aviation Agency (FAA), more than thirty-two major airports may be congested in the next decade and domestic air travel trips of less than 500 miles are more expensive than longer flights. HSRT can compete effectively with at least 50% of these short domestic flights, requiring less time and cost. Moreover, roads and bridges leading to airports are in such a sorry state that they will require several trillion dollars to repair. Alternatively, HSRTs can decongest roads, save energy, and reduce air pollution. If we take the Tokyo–Osaka model of the bullet train, the payoff period may be less than two years. But the new rail system will need the same access to tax-exempt bonds as highway and airport projects enjoy because of very high initial costs and also because the current road and air infrastructures cannot reduce fuel consumption, pollution, and oil imports unlike the new rail infrastructure. According to the HSRT technology advisory group, HSRT will use one-half the thermal energy of autos and about a quarter as much as airplanes per passenger. HSRTs are not affected by weather fluctuations and their land requirements are also less than the highway infrastructure per passenger. Trails of HSRT will run along the interstate highways with many stations for passenger drop-offs like highway ramps – opening an opportunity for rural economic development.

If we accept the above hypothesis of future dual track transportation system, we can now talk about corresponding parking changes: hard change and soft change (strategically) that will lead to a smart and combined transportation service paradigm embedded into RWPTTP given that high-speed rail systems are definitely on the horizon.

Hard Changes in Parking

National coordination of all parking spaces for accountability, elimination of all meters and gates, redistribution of spaces (not necessarily

---

5 Transport Reseach Council,, *In Pursuit of Speed: New Options for Intercity Passenger Transport, A Report published by the National Academies Press of Science Engineering Medicine.* Washington, DC: National Research Council, 1991.

one to one) as and where required in the dual track systems from the city business center, coordination of trucks' parking with weighing stations, elimination of heavy trucks in the city during commuting hours, use of electric and autonomous vehicles for transporting goods and people door to door, elimination of meter maids, no field trip for cash collection, complete separation of non-traffic violations from traffic violations, etc. The list is not exhaustive and can be a place holder to add later as they are seen appropriate.

Soft Change in Parking

Software emulation of all parking hardware functions, self-identification of vehicles and drivers as soon as they drive to a parking space, self-identification of all parking spaces, a national database of all parking spaces, a national database of all available parking spaces at an instant, one-click reservation of an available parking space closest to your destination, online auto monitoring of enforcement, no ticket and fines but auto step-up rates, pay-as-you-use rate, instant new service introduction for increased revenues, bringing civility with modern regulation, etc. The list is not exhaustive but with AI, cloud computing, and big data analytics it can go much further than is currently imagined.

Once the above concepts are well understood, smart and transportation service implementation where total automation is a focus will be feasible. Such a smart service will eliminate the subtle barriers of parking and transportation services from the operational, monitoring, and management point of views. Some skeptics may think it is just a dream but what is a dream today may be a reality tomorrow. Technologies may be our savior. Let us look at the state of available and on the horizon technologies. We need an affordable parking and transmission act (APTA), a vision of the future (out of box thinking). Such an act will help to allocate the necessary and needed funds from the federal government to get the project off the ground. No socialism or distribution of wealth can stop us heading for disaster. Only science and innovation are the answer.

EXAMPLES OF OTHER SERVICE INDUSTRY AND TECHNOLOGY LANDSCAPE

Three industries are worth mentioning in this context. They are the telephone company, the airline company, and the merchandise delivery company (Amazon). You can make almost free telephone calls using Skype, Vonage, WhatsApp, facetime, etc. You can have home-printed boarding pass, self-check-in, etc. Amazon uses a web portal and drones for merchandise delivery. Google is introducing autonomous vehicle (driverless driving and parking) for the last mile of transport. All these were possible because of different combinations of software and hardware technologies. But smart software was the key. The voice over IP enabled sharing backbone telephone resources via packet switching concept reduced the price of voice calls; smart kiosks and mobile software apps sharing workload with airline passengers contained airline costs; drones brought the distance and time down for the delivery of merchandise. With the emergence of AI, iCloud, big data analytics, and other technologies, the future looks much brighter. For example, airless tires (science matters), kill switch idea (making phone or autonomous vehicle inactive when stolen), and electric vehicles (gasless) will raise the transportation industry to the next level. Designing rooftop parking spaces for drones is not out of question.[6] What else can we achieve with latest technological tools if RWPTTP is implemented as the smart transportation platform? Priceline.com and airlines have their websites advertising their services on a regular basis .

In contrast, Google and other websites show negative images of parking as shown in Figures 9.3 (a) and (b). There is no shortage of them. Even many activists are playing little Robin Hood in different cities (New Hampshire and Texas) by volunteering meter payments when it expires. Hopefully, RWPTTP, based on open source products with plug and play tools, will bring advertisement tools for transportation services and provide a civilized image to the parking service and transportation services. I feel technology is ripe for the transformation of transportation services. And many are convinced that, with such a paradigm, customer service will improve even with the lower costs, with an advertisement platform producing extra revenues as other social media do. In the end, RWPTTP can be a repository of education, training, and other information like Google but exclusively for transportation services. Figure 9.4 gives a cost comparison based on a

---

6 IPMI conference theme – ignite session presentation.

FIGURE 9.3 Negative images of parking: (a) when a vehicle is clamped; (b) when poor security allows an assault and robbery to take place.

demo model in the lab. Preliminary results show the following improvement:

- Weighted savings per space is over $222 with no gates/arm

- Weighted savings with simplified arm per space is $131

- These savings include additional costs of enforcement for no gate option

Further refinements are possible when a full smart transportation service model following guidelines of RWPTTP for a dual track autonomous transport system is introduced.

VISION ZERO – A POLICY FOR TRAFFIC SAFETY

'Vision Zero' is a traffic safety policy that originated in Sweden and is being followed by many cities and countries. It is mostly hard change in the design of roads and streets that will impact parking and road transportation.[7] The author has decided not to elaborate much on the subject because RWPTTP, which is a software-based solution, will not be impacted by adopting a 'Vision Zero' policy. The whole philosophy of the vision is to improve road safety with less accidents and fatalities. Of course, in the age of autonomous

---

7 Roger Johansson, 'Vision Zero – Implementing a Policy for Traffic Safety', *Safety Science* 47, July 2009, Road Safety Division, Swedish Road Administration, Roda Vagen 78187.

Costing Items (weighted)	Gated / Space (weighted) in $	Registered Sin / Space in $		Savings / Space in $	
		RFID/Tag	RWPP	RFID/Tag	RWPP
Entry stn	40	20	0	20	40
Exit Stn	62	25	0	37	62
Cashiering or Pay on Foot	35	35	0	0	35
Gate & Arm	9	9	0		9
AVI -IN	-	4	0	-4	
AVI - Out	-	5	0	-5	
Hanheld Device (wghtd)	-	-	2		-2
RFID Reader (wghtd)	-	11	0	-11	
Signage	-	2	2	-2	-2
Enforcement	-	-	50		-50
Yearly Mntce	75	14	10	61	65
Paint	-	-	5		-5
Regstn S/W	-	15	0	-15	
Guaranteed S/W	-		10		-10
Systems Intgn	100	50	20	50	80
Total	321	190	89	131	222

FIGURE 9.4   Another cost comparison of smart transportation service (RWPTTP).

driving and autonomous parking, new road and parking facility design will play a vital role and must be accounted for.[8]

---

8 Craig Milligan and Rebecca Peterniak, 'Vision Zero: Principles and Checklist for Effective Adoption', 25th CARSP Conference, Ottawa, ONT, May 27–30, 2015.. Further data from Wikipedia.

CHAPTER 10

Why a New Business Model?

INTRODUCTION

The traditional parking business model was very simple. You can divide the parking scenario into three categories such as on-street, off-street, and garage (open lot and structured). The goal of each category was to earn revenues and make profits under tight control of local government regulations. A mechanism (street meter) or barrier (gates) were installed to control access. Customer service and convenience were secondary. Different financial parameters for the three business models are shown in Table 10.1 to 10.3.

Table 10.1 identifies appropriate items to configure an on-street parking business to earn revenues. Depending on the profitability objectives (10% to 30%), you can have different rates at different times of the day including weekends. If you want to be aggressive you can hire some enforcement officers to write tickets. You can have a strategy to impose fines and other penalties if tickets are not paid within the prescribed time periods. It is city's prerogative to play hard ball or soft ball with the following variables:

1) 24/7, or five or six days' operations

2) Maximum, minimum, or variable hourly rate

3) Enforcement policy and the ticket value for violation

TABLE 10.1 On-Street Financial Model

Configuration Items	Range	Cost	Comment
Spaces	1000	$1,341,000	Land, construction, and maintenance costs[a]
Capital for meters		$1,000,000	$1000 per space per meter
Interest annually	7%	$163,870	Borrowed amount
Insurance for the loan		$327,740	A factor of 2 – debt coverage
Manual operating cost	Four attendants @ $30,000 - per year salary & benefits	$120,000	Supporting two shifts: subtotal per year cost -= $1,611,610
	Other misc. expenses	$241,742	15% of the above sub-total – cash collections, credit card processing' and other miscella-neous energy expenses
Subtotal annual cost		**$1,853,352**	**Minimum revenues to cover costs**
Minimum total annual revenue required		$2,038,707	10% profitability = $185,335
Ticket revenues per meter per day requirement	$2,038/200 = $10.19	Only 5% of 50% ticketed and 2.5% fined pays the ticket[b] Four ticket hours based on four cars turnover per day beyond per-mitted hours	Assuming no pilferage and eight hours opera-tion at full capacity and 200 days
Scope of revenues	(Normal revenues per meter per day @ $5)	$0.5	Four ticket hours based on four cars turnover per day beyond per-mitted hours. For four turnovers: $5 × 4 = $20 per meter per day for regular fees. If only 5% pays it becomes $0.5 per meter per day

(*Continued*)

TABLE 10.1 (Cont.)

Configuration Items	Range	Cost	Comment
			with 200 days operation. Doubling (10%) the normal payment rate the revenue = $1. The goal should be 100% = $10.
Fine revenues per meter per day @ $20	$1	2.5% are fined @ $20 and 1.25% are collected. Four ticket hours based on four cars turnover per day beyond permitted hours.	200 days operations. 365 operation days are also possible. 2.5% are fined @ $20 and 1.25% are collected. Four ticket hours based on four cars turnover per day beyond permitted hours. With four turnovers potential revenue = $20 × 4 = $80. If only 2.5% pays fines, potential revenue= $80 × 2.5% = $2.
Total possible revenue outlook per day per space	$1.5	Aggressive enforcement and fines and penalties may produce more revenue. If 100% of 500 parked pay the normal fees the revenue will double from $0.5 to $1. If 50% are fined the potential revenue with four turnovers, revenue = $20 × 4. With 50% collection, revenue = $80 × 50% = $40.	

a Victoria Transport Policy Institute, 'Transportation Cost and Benefit Analysis II – Parking Costs' (www.vtpi.org).

b Aaron Adiv and Wanzhi Wang', *On-Street Parking Meter Behavior*. Ann Arbor, MI: University of Michigan Report, January, 1987.

4) Escalation of fines for non-payment within certain time

5) Court costs for contested tickets

The above configuration has inherent limitations of optimum performance due to:

- Rigid configuration
- Unfriendly and non-efficient enforcement process
- Expensive cash collections from underutilized meters
- New revenue base will need new meter
- Request For Proposal (RFP) process delays new installations
- Backend processing being vendor-specific, not always plug and play

Table 10.2 identifies items to configure a surface lot either owned by the city or the private owner. Similar to on-street parking, there may be some variations of the rate structure with a timetable for increasing revenues within 10% to 30%. The city may face some competition depending on the occupancy of city lots. In most cases, both are occupied more than 90% during busy office hours. The possibility of a parking violation is rare but penalties could be imposed for overstay or an abandoned vehicle. The city may have the prerogative to increase the rate and the private owner may follow suit within certain variables. A revenue model with different parameters may be formulated for the revenue and tradeoff analysis.

1) 24/7, or five or six days' operation

2) Different revenue schemes – reserved, daily, hourly, and monthly

3) Rates for each scheme for optimum payoffs

4) Exceptional violation

5) Court costs if tickets are contested

As in on-street financial model, the configuration has inherent limitations of optimum performance due to:

TABLE 10.2 Urban Surface Lot Financial Model

Configuration Items	Range	Hard Cost	Comment
Spaces	600		A typical surface lot
Capital for 600 spaces		$1,237,200	Land, development, and maintenance[a]
Capital for PARCS	Per structure of 600 spaces	$242,858	California City council bid report[b]
	Yearly maintenance	$48,572	20% of PARCS cost
Interest annually	7%	$103,604	Borrowed amount
Insurance for the loan		$214,208	A factor of 2 – debt coverage
Manual operating cost	Four attendants @ $30,000 per year salary and benefits	$120,000	Data – a report by Hunter Interest Inc.[c]
	One manager @40,000	$40,000	
Subtotal per year cost		$526,384	
	Other misc. expenses	$105,277	20% of the above subtotal for energy, payment processing, etc.
Total annual cost		$631,661	Minimum revenues
Minimum total annual revenue		$694,827	Assume 10% profitability
Minimum annual per space revenues requirement		$11,580	Assuming no pilferage
Daily revenue per space requirement		$57.9	Assuming 200 days operations
Other scope for daily revenue	Monthly pass per day per space @ $200 per month	$6.7	60 spaces are monthly pass holder for steady revenue.
	Short term parker @ $3 per hour per day per space	$24	120 spaces – turning over every 2 hours for a total of 4 turnovers.
		$10	

(*Continued*)

TABLE 10.2 (Cont.)

Configuration Items	Range	Hard Cost	Comment
	Daily parker @ $10 per day per space		The remaining 420 spaces are at flat daily rate.
Total possible revenue outlook	**Per space par day**	**$40.7**	**Rate variations are possible**

a Victoria Transport Policy Institute, 'Transportation Cost and Benefit Analysis II – Parking Costs' (www.vtpi.org).
b California City, 'A City Council Report in California While Evaluating WPS Bid', August 27, 2013. www.californiacity-ca.gov/CC/index.php/building/rfp-and-bids
c Andrew Harper, 'Financial Analysis of the Proposed Third/Second Street Parking Garage', Hunter Interests Inc., submitted to Santa Monica City, 2016.

- Vendor-specific technology with delayed software upgrades

- Limited service scope unless incremental and expensive technology is integrated

- Incremental technology integration becoming expensive and inefficient

- Long line at the access and the exit causing traffic congestion

- No rationale for rate increase

- Subject to revenue pilferage

- PARCS system grows by leaps based on the limit of spaces served by one system

- Not web enabled, resulting in expensive reconfiguration, and vendor-dependent

For structured garage parking, a new cost element of the structure must be factored in to calculate the revenue requirement. It is not considered here because it does not add any overall additional point other than higher revenue requirement. But the two models reflect the fact that revenue requirement of garage parking is higher than the street parking. Again, every time PARCS capacity exceeds the first limit level there is a jump of costs per space before the next capacity limit for more than one PARCS unit. Incremental technologies are being injected into the above models to improve some customer service. Two technologies stand out: mobile payment for on-street parking and credit card access and exit for off-street parking. Both impact the business model cost with little benefit either to the owner or the end user and add delay in processing the access and the exit.

Table 10.3 shows financial items of web-enabled application called soft meter application (SMA™) developed by Eximsoft International. During the initial research, early 2000, making parking services into web-enabled services was a challenge. We succeeded in that challenge and proved that it can be done even with existing hardware constraints. There were a few objectives for such an application related to customer service such as:

- Customer registration for premium service
- Combining on-street and off-street parking under one application simplifying backend processing
- Improved customer service providing reserved and guaranteed parking space for off-street parking
- Day long pass for on-street parking
- Use of space # and block # to activate parking application using smartphone
- Make changes of parking reservation as and when needed
- Accommodation of autonomous vehicles

Operators or owners in airports or similar locations with such application saw an increased revenue due to a convenient parking service to end users. It is easy to reconfigure parking spaces with a small change in software codes. SMA™ does not need any hardware in the field and it is easy to generate new revenues without field hardware – only a space identification is required in smartphone apps. The SMA™ configuration favors digital technologies such as drone, software modules, and photo voltaic (PV) cells. Accommodation of 50% to 62% more autonomous vehicles in 1600 spaces of the previous two models is a plus. Capital costs and debt coverage for capital are higher with modern technologies: $1,267,254. With 20% profitability, your expected yearly normalized revenues must be over $1,520.705 or $587 per space per year, which is lower than the combined two previous models. With this universal framework one can be innovative to

TABLE 10.3 SMA™ Financial Model

Configuration Items	Range	Cost	Comment
Spaces (Table 10.1 + Table 10.2)	2,560		1.6% autonomous cars[a]
Capital for configuring 2,560 spaces		$1,237,200 + $1,341,000 = $2,578,200	Land, construction, and maintenance of space[b]
Web development capital	Infrastructure cost including operation management	$500,000	Eximsoft's large-scale software development report
Marking space # and block #	One-time	$16,000	For open lot and on-street spaces
Capital for PVs	600 lot spaces	$3,000,000	Per MW average solar cost is $2,500,000[d]
Energy credits due to PVs	60%	-$1,800,000	600 spaces generating 2.7 MW[e]
Subtotal capital cost		$4,294,200	
Interest annually	7%	$300,594	
Insurance premium		$601,188	A factor of 2 – debt coverage
Operating costs	Two attendants @ $30,000 per year salary and benefits	$60,000	Automation reducing manpower
	1 manager @40,000	$40,000	
Subtotal per year cost		**$1,001,782**	
IT and other maintenance costs		$150,267	Assuming 15% of per year capital cost – more savings when staff trained
Total per year cost		**$1,152,049**	
	Other misc. expenses	$115,205	10% of the above subtotal
Grand total annual cost		**$1,267,254**	Minimum revenues to cover cost
Minimum annual revenue requirement		$1,393,979	Assume 10% Profitability
Minimum annual per space revenues		$545	Revenues via automated invoices – less human error

(*Continued*)

TABLE 10.3 (Cont.)

Configuration Items	Range	Cost	Comment
Minimum per day revenue requirement		$2	24/7 and 300 days operations due to – automation
Other scopes of daily revenues per space	Step-up rates for first two hours @ $5/hour	$20	50% violators pay extra without ticket and fine over normal rates -possible by automation. Four vehicle turnovers assumed.
	Step-up rates for the second two hours @ $10/hour	$10	No collection agency for fines. 25% assumed to stay beyond four hours violations. Four vehicle turnovers assumed.
Daily revenues per space potentials		$30	Many creative schemes possible for increasing further new revenue generation

a Pete Bigelow, 'A Big Makeover is Coming to the Parking Garage of the Future – Thanks to Autonomy', *NYT*, July 26, 2016.

b Victoria Transport Policy Institute, 'Transportation Cost and Benefit Analysis II – Parking Costs' (www.vtpi.org).

c Amalendu Chatterjee, 'Modernizing Parking Access and Revenue Control System (PARCS) Using Parking Web Portal (PWP) for Cost Effectiveness' in *Operation Efficiencyd, Uniformity and Increased Business Margin*. Eximsoft's Internal Business Strategy Report, March 2004.

d 'What is the Average Cost of Solar Panels in the U.S.?', www.thestreet..com / technology/average-cost-of-solar-panel (2019).

e 'Lockheed Martin Parking Catches Sun Power', press release, January 26, 2016.

optimize revenue generation and tradeoff analysis. Advantages of the new web parking business model:

- On-street, garage, and surface lot spaces under one application

- Forward-looking services with reduced lot space per vehicle

- Improved and efficient business process

- Adding and deleting spaces with simple software codes

- Decreased cost per space as spaces are added, reversing the current trend because of hardware independence

- Vendor-independent technology to define new web-based customized services

With the new technologies, the cost trend may be reversed because of elimination of vendor-specific upgrades including labor-intensive activities. PVs will reduce energy cost requirements by 50% to 60%. Software emulation of the hardware model may improve scalability compared to the existing hardware model. An example has been worked out to disprove the general belief that digital technologies are expensive under some cost assumptions and the setup stage of the RWPTTP roll out. Overall benefits outweigh costs. Digital technologies open up the scope for new revenue. Autonomous vehicles (cars or trucks) will require autonomous parking and integrated transportation systems. Such autonomous business models may be complex but we have to be ready. The industry will become virtual to be on a par with Uber/Lyft like services.

The above configurations and business strategies with almost century-old technologies did not address the real issue of parking and transportation services discussed in the other chapters of this book. Of course, solar PVs and web services have added an improvement to the bottom line. Is it enough? No, we need to do more. In USA, the congestion problem, city by city, is well documented. Efforts are being made to transfer such congestion to the neighboring arteries. That is why the current business model of making money or providing services to the end user under the current configuration is not sufficient because it transfers the problem from the core city area to other areas. It is important to note that we need to be careful about transferring one city's congestion to neighboring highways leading to other cities. In this context, we need to develop strategy for a global approach to an integrated solution. For example, on-street and off-street parking are artificial partitions, adding to many of the city's unnecessary problems of discrimination. In addition, piecemeal solutions can exacerbate problems in other parts of the system. Solving both problems may, for example, lead to congested highway commutes. In many cases, it may be concluded that the current road infrastructure needs a backup system such as light rail. I call it the dual track system and if we can coordinate parking as part of this dual track system we will go somewhere – some of this is already happening.[1] Along these dual track systems, parking spaces can be relocated to a convenient places for the last mile commuting distance to relieve city congestion and also have

---

1 For example, the Los Angeles to San Francisco SpaceX system.

a global view of all spaces – the precursor of RWPTTP. Technology is on our side and it is a question to choosing the right combination for a total virtualization with stability. The new business model will need a total re-engineering of backend processing to improve performance and service delivery with a global view. In addition, the new model must collect real-time management information to handle many online requests for permits, changes, and payments.

RWPTTP Business Model

Out of the three models above, the financial model of SMA™ shows a technology edge with customer service including consolidation of on-street and off-street parking. The financial model addressed the local individual operator or owner model only. The challenge for modern parking services has now reached a different level. We need to go beyond web-enabled services for a global model with a standard for a national uniformity. A full-scale financial model may not have been developed yet but I have outlined its possible extent – a Google-like parking application for the nation. As stated earlier we need an umbrella parking business model to address the needs of smaller parking owner (private and public) as well as the large parking owner.

We need to develop the RWPTTP financial model following the connectivity transportation model of Figure 10.1 – a final destiny of the next century. The characteristics of future parking needs must be envisioned and technologies that could impact such a vision must be understood.

Visionary characteristics should be as follows:

- Raising the parking service to a national level for convenience and experience.

- Making cities free from traffic congestion and pollution.

- Reducing total parking spaces proportional to the number of vehicles (1:1 at least).

- Building no garages for parking – making all open to take advantage of renewable energy.

- Addressing city congestion by bringing parking traffic and transportation ogether.

FIGURE 10.1 Autonomous (RWPTP) parking – technology connectivity outlook.

- Strongly coordinating public and private services for a national consolidation.

- Broadcasting space availability for compulsory reservation to reduce GHG emissions and traffic congestion.

- Exploring parking spaces for renewable energy using photo voltaic (PV) cells.

- Automated step-up rates instead of manual tickets, fines, and escalation of fines.

- Defining suburban parking district (SPD) away from the busy city business district (CBD), with frequent transportation services between SPD and CBD.

- Public and private cooperation for virtual transportation services as envisioned by Uber/Lyft.

- Promoting dual track transportation (light rail and roads) and relocation of all parking spaces along such tracks.

- Better handling of workload partitioning following airline models such as check-in, baggage handling, ticket purchase.

- Reducing vehicle ownership and introducing autonomous vehicles for transporting people and goods in parallel.

- Stopping trucks from entering cities, instead provisioning more truck parking along interstate highway corridors and weighing stations.

- Pre-checking of all truck contents including maintenance records and driver's driving habits along its route.

- Offering courses at community-level colleges to raise awareness and developing manpower resources for the next-century parking and transportation paradigm

These, of course, are lofty goals. Achievability will depend on the maturity of the right technology. Examples (not exhaustive) of possible technologies that need to be integrated may be as follows:

- Information technology (IT) for improved backend processing and auto invoicing

- Drones for auto identification of space, vehicle, and end user

- Artificial intelligence (AI) for smart monitoring of the system and partitioning individual roles

- Smart software to contain costs for new service instantaneously, following the model of free telecommunications services (e.g. Skype).

- Cloud computing for ongoing and instant decisions based on data collection and its analysis.

- Software emulation of hardware elements to contain costs.

- Autonomous vehicle technology for enhanced driving, parking, and transportation of people and goods.

- Electric vehicle technology to reduce pollution and linking other connectivity technologies for an improved coordination of all services.

- Mobility and availability of open space as and when needed.

- Single billing and payment or virtual payment (blockchain) system.

- Improved security of parking space, such as the trusted parker program (TPP™) for securing parking areas.

- Parquest™ for broadcasting availability and a direction to get there (a Mapquest parallel).

It is too early to identify all cost items of the above proposed business model. Hopefully, a later edition of the book will provide additional details based on outlines now known or that may develop later as a result of further research. The good news is that all benefits of such an outreach will outweigh the cost of technology. In addition, it will be possible to get rid of undesirable remnants of the old system such as unfriendly enforcement policies, unfavorable regulation, and use of judicial system resources for traffic violations. If all strategies and technologies are well orchestrated we can rationalize the new business model to accommodate a total autonomous-based transportation ser-vices – hardware and vendor independence, dignity, civility, convenience, instant and automatic service by kiosk and smartphone, mobility, con-nectivity of devices, workload sharing, and much more. The enabling concept, Robust Web Parking Portal (RWPTTP), is to establish a network of connected vehicles, electric vehicles, GPS, BLE, software, software emulation of parking hardware functions, AI, parking database, vehicle identification number (VIN), sensor devices, object oriented program-ming, automatic vehicle identification (AVI), license plate recognition (LPR) and correlation among VIN and related LPR. The RWPTTP infrastructure has been described in Chapter 3. The objective is to leapfrog automation of parking features and functions. The following perspectives of software algorithms are further expanded relating to RWPTTP:

- **Parking Database** – Department of Motor Vehicles (DMV) tracks owners by a database with certain attributes: owner, address, county, vehicle identification number (VIN), license plate (LP), and other relevant information for property taxes, emission tests, vehicle transfer, and other tags. Similar attributes, such as space number, street loca-tion, city, state, owner of the facility, on-street/off-street characteristics, occupied or empty, etc. are being proposed for parking space to relate to vehicles for parking automation.

- **Object Oriented Programming** – A software programming tool where attributes of different databases act as useful information to write instant applications for new services and new revenues. Integration of all parking options on a mobile Internet platform with different Application Programming Interfaces (APIs).

- **Correlation of Automation Attributes** – Attributes of online parking automation are: online finding of parking spaces, knowing the parking rate, the proximity, notification of illegal parking, marketing/promoting services, auto billing, knowing the owner when parked, correlation of trucks and weighing stations for route selection, highway regulations, mitigating commuter and truck driving with well-designed truck parking along highway corridors – Vision Zero implementation.

- **Photovoltaic (PV) Cells as Renewable Energy and Sustainability** – Charging of electric vehicles while parked will add one more dimension to the new business model. All energy needed for ITS and electric vehicles could be sourced from PVs covering lot spaces. An analysis of it contends that open parking spaces in USA as renewable source may be equivalent to the total of the present grid connection (see Chapter 6).

- **Drone Technology for Parking Enforcement and Traffic Management** – Drone technology for parking and traffic are two promising applications. Drone is one common technology unique for both traffic and parking management for integrated ITS and parking operations.

- **Vehicle Connectivity for Traffic and Parking Updates** – Connected car to be equipped with Internet access via wireless network or other smart devices inside or outside the vehicle – vehicle to vehicle (V2V) and vehicle to infrastructure (V2I). Both will improve the safety, security, and traffic congestion of driving and parking.

- **A Suburban Parking District (SPD)** – SPD may be defined as city's transportation hub away from the city's CBC with frequent virtual transportation systems between them – decongest cities from cruising, resulting in reduced GHG emissions. Private and public transportation systems including autonomous vehicles will compete for revenues between these two points of interest – virtual and intelligent transportation system. A parallel is the Truck Parking District (TPD) along interstate highway corridors to coordinate weighing station resources and to combat commuter traffic on highways during office hours.

- **Virtualization of Future Services** – Voice over Internet (VOIP) which not only contained the cost of long-distance telephone calls for decades but also added multi-media capabilities establishing a precedent (e.g. Vonage and Skype). Virtualization will also reduce the manual process and improve efficiency.

- **Coordinated Management** – Centralized parking operations (CPO) mimicking the traffic management center (TMC)

Incremental efforts to modernize segmented parking installations are not sustainable because we need a global view of all the space resources. An incremental approach would cost more if an overhaul is needed later to deliver the potential benefit. Technologies are available for the visionary solution proposed here. it will be a failure on our part if we cannot improve things for the next generation. The RWPTTP infrastructure needs to be populated with all the hardware and software modules of parking and traffic functions. These modules will be glued together with application program interfaces (API) to develop a wide range of customer applications and services. Hopefully, optimal implementation will reflect the true cost of every technology element proposed in the new business model – with no government subsidies by third parties unlike today. Rates derived from true costs will let everybody pay his/her share of the parking proportional to the usage independent of location – on-street or off-street parking. The Google-like Parking Infrastructure will also treat all stakeholders equally. For example, even small parking lot owners with five to ten spaces will be able to market them whenever they feel like connecting to the web and will be paid based on their usage – like car sharing (Uber/Lyft) or room sharing (AirBNB). In addition, parking services need to be automated for autonomous vehicles, following models of automated toll and traffic services using the latest technologies. Integrated public transportation infrastructure will automate the invoice generation and payment process for all highway toll, parking, and traffic charges and violations. The objective is to change the long-standing perception of parking as a money-grabbing proposition for the great convenience of public transportation services. The above scheme may be complex and expensive. The objective of this chapter has been to compare current cost elements and relevant technology perspectives with that of what we should do in future:

1) Generate interest and develop expectations;

2) Explain achievable results as we make progress;

3) Establish a better framework to move forward.

Index

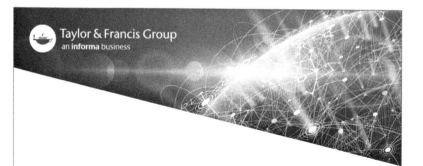

Milton Keynes UK
Ingram Content Group UK Ltd.
UKHW031532071024
449327UK00005B/102